New Age of Robotics and Modern Computer Vision

Advances, Innovations and Applications

New Age of Robotics and Modern Computer Vision

Advances, Innovations and Applications

Editors
Manisha Vohra
Prakash J.

Central West Publishing

This edition has been published by Central West Publishing, Australia
© 2024 Central West Publishing

All rights reserved. No part of this volume may be reproduced, copied, stored, or transmitted, in any form or by any means, electronic, photocopying, recording, or otherwise. Permission requests for reuse can be sent to editor@centralwestpublishing.com

For more information about the books published by Central West Publishing, please visit https://centralwestpublishing.com

Disclaimer
Every effort has been made by the publisher, editors and authors while preparing this book, however, no warranties are made regarding the accuracy and completeness of the content. The publisher, editors and authors disclaim without any limitation all warranties as well as any implied warranties about sales, along with fitness of the content for a particular purpose. Citation of any website and other information sources does not mean any endorsement from the publisher, editors and authors. For ascertaining the suitability of the contents contained herein for a particular lab or commercial use, consultation with the subject expert is needed. In addition, while using the information and methods contained herein, the practitioners and researchers need to be mindful for their own safety, along with the safety of others, including the professional parties and premises for whom they have professional responsibility. To the fullest extent of law, the publisher, editors and authors are not liable in all circumstances (special, incidental, and consequential) for any injury and/or damage to persons and property, along with any potential loss of profit and other commercial damages due to the use of any methods, products, guidelines, procedures contained in the material herein.

 A catalogue record for this book is available from the National Library of Australia

ISBN (print): 978-1-922617-59-0

Preface

Robotics was previously considered only for limited activities and applications and was suitable only in certain kind of applications. However, the recent few years has seen the evolvement of the robotics which has resulted in bringing about New Age Robotics and it has shown some remarkable achievements. For example, now robotics can be successfully used in complex applications like assisting in healthcare sector for carrying out surgeries, etc. In times like pandemic where the situation makes it difficult to work, robotics comes across as just the need of the hour. In the recent pandemic, the evolved new age robotics played a vital role. It was efficiently being utilized in various hospitals to give medicines to patients, to provide them with food, etc. Robotics is also efficiently used in agriculture, industrial and various other areas. Computer Vision helps in understanding the visual images and videos. Computer vision in the recent times has seen a lot of growth. It has evolved and is quite modern and useful. It can be used in different applications. For example, computer vision is highly beneficial when applied in healthcare sector, in monitoring systems, in self-driving vehicles, etc. There are major advances and innovations taking place in robotics as well as in computer vision.

This book titled, **New Age of Robotics and Modern Computer Vision: Advances, Innovations and Applications**, is a collection of book chapters explaining the different aspects related to new age robotics and modern computer vision which help in understanding the innovations and advances taking place in them. It also focuses on explaining the various applications of new age robotics and modern computer vision in different areas.

Chapter 1

The aim of this work is to reflect intensely on questions from a philosophical and epistemological point of view about what vision and perception are within the concept of evolutionary robotics. In this sense, it is essential to understand what vision and perception are, from the point of view of human actions and behaviors, since robotic studies and research on the rise are concerned with an increasingly refined and complex degree in the sense of seeking develop robots that mimic and express essentially human characteristics. From the perspective of social robots, it is intended that interaction and participation in the daily life of human beings become increasingly present and common, as a natural and normal fact. However, many questions remain open, such as the values that should guide the strategies of relations between humans and machines in the near future. Would it be the case of trying to investigate the notion of a robotic epistemology of what vision, perception, curiosity, tenderness, affectivity and ethical values are in a society in which artificial technology will increasingly become a conditioning parameter in certain social relations, is a question that needs attention. Finally, the recent period of the coronavirus pandemic has aroused great interest in the use of robotic technology not only as a resource to help preserve life, health and well-being, but also to define new modes of social communication, sociability and partnership in different contexts. Thinking of social robots effectively as co-workers and company for affective support requires a great innovative vision and frequent parallelism with human actions, behaviors and values, especially if we wish to aspire to develop affective social relations with them like the ones we have with co-workers.

Chapter 2

Robotics has progressed a lot. There have been a lot of advances in robotics. Now there is new age of robotics which has many possibilities. It can be used in various sectors. In this book chapter, with the advances in robotics and the presence of new age of robotics, the application of robotics in different sectors like healthcare and agriculture will be discussed. Along with it, the need of robotics in these both sectors and advantages of application of robotics in these sectors will also be discussed in the chapter.

Chapter 3

Electromyography (EMG) signals can provide benefit to biomedical/clinical applications, etc. Muscle EMG signals through surface electrodes or needle electrodes, require advanced methods. In our work, we implement a hardware set up of the signal acquisition system and develop code to realise the servo actuation in real time to observe the muscle activation and the movement of the servos accordingly. We further analyse the acquired data in the time and the frequency domain in order to identify most prominent features which can be used to classify the various gestures of the wrist and the finger movement by using various signal processing techniques. These features are compared and verified by using data from existing databases online such as the Ninapro. A performance comparison study of various EMG signal analysis methods is also provided. This will help to further develop a powerful model for its application in the prosthetic arm. A 3D model has also been in the making that would house the hardware in future iterations of the project.

Chapter 4

Internet of Things (IoT) and robotic based system presented in this work is a system which works to build a smart environment. The system has enormous amount of data which is stored over the cloud. The framework consists of different sensors consisting devices working together to accomplish the task. The readings generated by the sensors are used by robots in its activity. The sensing devices are installed to collect vital signs of the patient and store it over the cloud. The robot is embedded with several algorithms to perform the task of giving medicine to the patient, etc. The robot can be monitored and the data obtained can be used for analyzing robot activities. It can be used to cross-verify their doings and measure the performance. If this system is faulty, then it could affect the patient's health. Hence there should be security measures applied. The system provides security to data by 4.84% and the faulty devices are detected at an early stage by 8.94% compared to the system without any security measures.

Chapter 5

Stress is a major problem in our society, as it is the source of many health problems. This work concentrates on this problem which confronts everyone today. Stress will have adverse effects on people if it is not detected at an early stage. In this work, human emotions with monitored facial expression sense stress as a function of the captured image. An approach based on convolutional neural network (CNN) was introduced. Here, the considerable emotions are happiness, sadness, frustration, surprise, disgust, apprehensive, neutral. Thus, deep learning has an ability to understand the characteristics that will allow machines to generate perception. The emotional analysis performed at the end of each iteration suggests that reducing the invasive nature of the device can influence user perceptions and improve classification performance.

Chapter 6

Cyber-physical systems (CPS) are growing at good pace. Their use in different applications has been increasing. Their demand and popularity is growing. In such case, keeping them securing them is important. It needs to be made sure that they are secure. In this chapter, an introduction of CPS will be provided. Along with it, how CPS can be kept secure using neural network and deep learning will also be discussed.

Chapter 7

Intelligent systems based on Artificial Intelligence (AI) and Machine Learning (ML) are poised to make disruptive and transformative advances in the biomedical domain. These systems assist in contexed-relevant data synthesis and automation in the biomedical industry. A formal definition of AI states 'the capability of a machine to imitate intelligent human behavior.' The AI-laden systems assist humans in modeling human reasoning to execute a problem or bypassing human logic and exclusively use a large volume of information to engender a solution. On the other hand, the AI system assimilates elements of human reasoning without accurate modeling of human processes. The framework of the AI system consists of two subdomains -methods and application. The Methods subdomains include evolutionary computing, expert systems, machine learning (ML), fuzzy systems, and probabilistic methods. The neural networks and support vector machines are two subdivisions of ML. The probabilistic methods are subdivided into the Bayesian networks and Hidden Markov models (HMM). The application domains include Natural Language Processing (NLP), predictive analytics, robotics, vision (Image recognition, machine vision), text-to-speech, and speech-to-text.

Symbolic AI is based on the high-level human-readable representation of problems and logic. ANNs are nowadays provide reliable results related to biomarker identification and classification of disease. With the help of artificial intelligence, it is possible to collect more information and samples on digital platforms, store data, and perform data analytics to pile up the affluence of data from biomedical databases such as genetic mapping on DNA sequences. The classification of genes and cancer cells, protein function, disease diagnosis, and disease treatment is more accurately assisted via AI implementation. Bio computation systems help improve various fields such as molecular medicine, drug development, and gene therapy. This chapter explores the potential of AI and ML in the biomedical domain.

Intelligent systems can be made using AI and ML for application in biomedical domain. This chapter covers an overview of machine learning, artificial neural networks (ANNs) and AI. Also, the application of AI and ML in the biomedical domain is discussed in the chapter.

Table of Contents

1	**Robot Epistemology: Vision, Perception and Curiosity as Innovative and Disruptive Challenges in the New Age Robotics** Paulo Quadros	1
2	**Advances in Robotics: Applications and Advantages** Manisha Vohra	15
3	**Prosthetic Arm for Rehabilitation Robotics** Aishwarya S, S. Srinivasan, Sushmitha M. and Veena Hegde	23
4	**Monitoring System for Patients and to Detect Faulty Devices** Ambika Nagaraj	53
5	**Early Stress Identification and Detection from Facial Expressions** Sheema Sadia, Apurva Kumari and M. C. Chinnaiah	71
6	**Keeping Cyber-Physical Systems Secure Using Neural Network and Deep Learning** Manisha Verma	79
7	**Intelligent Systems in Biomedical Domain: The Perils and Promise of Hope** Kalpana and Abhishek Maurya	87

1

Robot Epistemology: Vision, Perception and Curiosity as Innovative and Disruptive Challenges in the New Age Robotics

Paulo Quadros
School of Communications, University of São Paulo, São Paulo, Brazil

Abstract: The aim of this work is to reflect intensely on questions from a philosophical and epistemological point of view about what vision and perception are within the concept of evolutionary robotics. In this sense, it is essential to understand what vision and perception are, from the point of view of human actions and behaviors, since robotic studies and research on the rise are concerned with an increasingly refined and complex degree in the sense of seeking develop robots that mimic and express essentially human characteristics. From the perspective of social robots, it is intended that interaction and participation in the daily life of human beings become increasingly present and common, as a natural and normal fact. However, many questions remain open, such as the values that should guide the strategies of relations between humans and machines in the near future. Would it be the case of trying to investigate the notion of a robotic epistemology of what vision, perception, curiosity, tenderness, affectivity and ethical values are in a society in which artificial technology will increasingly become a conditioning parameter in certain social relations, is a question that needs attention. Finally, the recent period of the coronavirus pandemic has aroused great interest in the use of robotic technology not only as a resource to help preserve life, health and well-being, but also to define new modes of social communication, sociability and partnership in different contexts. Thinking of social robots effectively as co-workers and company for affective support requires a great innovative vision and frequent parallelism with human actions, behaviors and values, especially if we wish to aspire to develop affective social relations with them like the ones we have with co-workers.

1. Introduction

Undoubtedly, in last few decades, the development of robots has reached an unimaginable level of disruptive enhancements due to the improvements in artificial intelligence and neural networks, which have brought new perspectives for the redefinition of robotic visual perception design on a large scale.

However, there still remain the relevant issues regarding what we strictly define artificial vision and perception in comparison to natural vision and perception, applied to biological and human context.

From a philosophical and epistemological point of view, one can understand that the notion of vision and perception, in terms of robotics, continues to take their gradual steps of development, making the process of interaction and apprehension with everyday human reality even more complex.

Czech writer Karel Čapek used the word robot in a play. This word is derived from the Slavic term "robota," to designate artificially constructed servants who were to be relentlessly obedient to physical labor and, by extension, to the forced and even slave one.

Naturally, over time, the human being, observing the nature around him, learned to imitate it constantly, and to improve knowledge in the art of mimesis of nature, in a broad aspect. In this way, he will increase his power of intervention in nature, to create and improve resources for protection of own self, resources for comfort and survival.

This power of intervention involves the constitution of the first tools invented by human beings, so that they could look for food and protect themselves from possible predators. Later, after the development of language and its registration form, the first machines appear, which mimic the behavior of nature.

All machines, and by extension, robots that are sophisticatedly high-enhanced machines, are imitations and interpretations of nature and the web of life in general, that is, they are conceptual mediations or metaphors with high symbolic power for representing continuous standards of behaviors from nature and life.

Many technologies have recently been created both to increase man's power of extension and intervention in distant and austere environments, as well as to increase the perception of his senses and abilities (speech, writing, hearing, sight, smell, taste, touch).

This can be understood in the recent invention of various means of communication: books and newspapers (extension of writing), radio (extension of the ear), cinema and television (extension of the ear and sight), and internet (extension of the book, newspaper, radio, cinema, television, etc.).

In this sense, robots currently seem to symbolize the notion of extension not only of the senses in general, but of life itself, imitating the amplified and augmented abilities and complex behavior nature of humans and animals.

Supposedly, all robotic senses studied and developed in research laboratories are attempts to program machines with algorithms that repeatedly and artificially imitate different human and animal behaviors at the same time.

It is, therefore, the maximum extension of biomimetic elements, with an increasing power of autonomy and decision, aggregated in a single technological mechanism (common communicating vessel, in cybernetic terms).

1.2 Vision and Perception in the Human and in the Robotic Context

Perception can have a visual and social characteristic, for example, when capturing and interpreting visual stimuli to transform them into information, and when interpreting the behavior of others.

According to author in [7], perception is a phenomenon of apprehension of knowledge related to experience. Therefore, seeing and perceiving in detail is related to the accumulation of analyzed and contextualized information. [7] also adds that the visible is what is learned with the eyes, when looking closely, through the careful movement of the eyes, while the sensitive is what is learned by feeling.

In this way, this is how sensitive perception is developed, in the act of looking at something very carefully, realizing that what one is looking and feeling deeply is what one sees in oneself when looking outside.

What makes the human gaze a field of differentiated and singular intelligence is perceiving and discovering the forms and meanings of the visual object. And it is in this way that the link is established with the sense of beauty that human beings feel in relation to what they see, a possibility free from the valuing dynamics of the gaze.

Because, when the look becomes intelligent, it also establishes endless affective links of correlations that connect memory to the future, the concrete to the abstract, perception to concept, and it is able to retrace this path also by the inverse interpretive path.

In this way, visual intelligence is very restricted to the informational operationalization of visual data extracted from the environment to which the gaze or optical device is directed. And, on the other side, in this sense, the intelligent look not only mechanically collects visual information, but it interprets its sense analytically and critically, transforming it into something that goes beyond a simple and a merely objective capturing look, because it has a sensitive dimension of knowledge around us, perceiving and signifying it.

The human eye endowed with intelligence not only sees a mass of fragmented images, without apparent relationships, in a continuously passive way, since it elaborates a symbolic construction of the perceived reality.

In addition, the human gaze wanders over objects and scenarios, giving them a context of apprehension of the observed reality, and this is precisely what gives a sense of singular intelligence to it, that is, this continuous movement of seeing the meanings that are present in the seen image.

The creative gaze transforms the vision of what is seen at each moment, giving it new configurations of apprehension and understanding, as it invents and reinvents free possibilities in relation to what is observed and perceived (meaning and re-meaning in all senses).

It also reveals a condition of projecting aesthetic and ethical values, as a knowledge journey of dialogical-dialectical senses of outside and inner reality. And it is a look with a fruitive content, which problematizes and inventories questioning hypotheses at all times, as it has an investigative dimension, acting as a detective internalized in the mind of the one who looks.

Once one can understand that all robotic senses, including the field of visual perception, are dynamic enhanced biomimetic components of human and animal senses, such new technological mimesis end up redefining the new frontiers between the technological imagination and the real technologized principle in a concrete form of coexistence.

This means that several aspects of past and unforeseen re-imitations of nature and life, as well as conceptual re-analogies, and re-abstractions of specialized behaviors begin to be confronted and overcome, since certain patterns are becoming to be better understood in their level of refinement and possibilities of algorithmic implementations.

In this aspect, Wienerian cybernetics (also called first order cybernetics) [10], which is fundamentally based on the communication and control of animals and machines, and by extension, on human beings, becomes relevant in the idea of metaphors and biomimetic analogies of nature, generally for the development of non-human technologies.

In Wienerian terms, the concept of non-human technology differs entirely from basic human technology [10]. Basic human technology refers to what the human being is in its essence, represented by: the human brain and mind, the human body in general with all internal organs and limbs (hands, arms, feet legs), involving all abilities and specific skills acquired and developed over time.

On the other hand, non-human technologies are precisely those that come from a process of cybernetization, artificialization, robotization, cyborgization, androidization, and so on, always initially seeking to establish analogies and symbolic metaphors with the human, to gradually go breaking with pre-established and delimiting dogmas in the field of distinctive autonomous behavior.

Along with Wiener's cybernetics, prostheses are not only substitute instruments for biological intervention, such as: arms, legs, teeth, for example, but they can be understood as any other elements coupled to the human body and their senses to fulfill a similar function or amplifier in exterior and independent environments.

And within this complex perception, all technologies can be understood as prostheses, such as: language, telephone, radio, television, internet, neural networks, artificial intelligence, and robotics, not necessarily being coupled to the body human, but representing its power of magnitude realization in terms of extension, embodiment and transcendence, establishing other fields of symbolic materiality or radical immateriality.

[6] makes use of the Wienerian idea to understand the various technological media as powerful ways to enhance human senses. In this sense, aren't robot prostheses a way to expand the human power to intervene in the world and in nature, its field of vision and perception that mimics the human gaze and understanding of the surrounding environment?

On the other hand, the so-called second-order cybernetics, improved by authors in [5] deals with the issue of complexity and diversity in the web of life, and its relationship with artificial systems, neural networks, computing and robotics.

In this sense, these scientists conceptualize the view of autopoiesis (self-creation, self-generation idea), in order to establish analogies with self-learning algorithms and autonomous mechanical systems, taking a significantly new evolutionary step in the development of highly complex learning machines such as robots.

1.3 Robotic Curiosity versus Human Curiosity

For [1], curiosity is an element of the evolution of life, employed at different levels, "ranging from the molecular scope to the social spheres of culture", leading to the perceptive hypothesis that something of new and unprecedented is emerging, which demands a multi-referential way of thinking.

[1] also argues that brain evolution is associated with predictive interactions that living beings develop when moving through their environment, since they needed to learn to anticipate the result of each movement performed. This makes living beings improve their pre-assessment and post-assessment processes, to analyze and interpret the mobility process from the analysis of sensory data that they receive and actively transform into something perceived with meaning.

In addition, the author in [1] believes that curiosity is the basis for the interconnectedness with emergence of new forms of innovative thinking and acting, considering the AI and Robotics influences of how interpreting curiosity. That is what the author in [1] calls creative curiosity, which is based in the correlation with arts means the creation of the new. In this way, the author in [1] also defines the concept of curiosandi i.e., curious to defend curiosity as a sense of enjoyment (fruition) as an innovative, entrepreneurial and libertarian spirit, which leads to the issues of robot and machines singularity, as advocated by [8], as the next step of artificial intelligence improvements.

In this way, every perceptual action corresponds to a predictive dimension, which guides the movement's action in relation to the environment, with no reaction that can be considered as an isolated part of the perceptual process. In other words, perception can be understood as a process of anticipatory construction, with a prospective dimension, in which the movement of the being streamlines the relationship of evaluating and predicting possibilities.

Nonetheless, [1] defines curiosity as the function of living beings in evaluating and predicting possibilities, considering that perception and prediction are interacting processes both in neurological activity and in the aspect of corporeality as a whole.

In this way, the senses capture the reality dimension, in a sensory and sensitive way, activating cognitive processes, which, for [1], would be the origin of the action of curiosity as an emergent phenomenon.

Still, in the field of human curiosity, [3] points to the difference between naive curiosity and epistemological curiosity. For [3], the first one resides in getting knowledge of pure experience and through common sense perception, while the second one focuses on criticality approach, by applying systematic use of rigorous methods of analysis, observation, concern, with the morphogenesis of knowledge, values and attitudes.

In this regard, [3] also warns that curiosity as "inquisitive restlessness" seeks to unveil something hidden as an integral part of the vital phenomenon of knowledge, since without curiosity there is no creativity, as it drives the human being to a permanent inquiring process of learning.

On the other side, according to [4], for robots to succeed in new tasks, an effort will be needed for them to learn how humans learn, interact and dialogue in the same way, with actions of autonomous initiatives.

Thus, the idea is that robots can interact more and more in open environments, with flexibility to perform dedicated tasks, learning to deal with unforeseen events, even if they are not linked to their original functional activity.

The interaction between humans and robots, in the field of robotic innovation, seems to be the future path of the field of disruptive technological development, since it is intended that robots start to interact, share and collaborate more actively with human beings, in a process of substantial improvement to quality human-robot interaction.

This means continuous research on human behavior and biomimetic capacity of artificial and robotic systems, which presupposes the development of robots not only specialists in certain complex activities, such as autonomous space systems or rescue robots.

The new social robots will face the challenge of increasingly refined social interaction, implying a more enhanced sensory field of vision and hearing, as well as symbolic filtering of ambiguous bodily and gestural meanings.

1.4 In Search of a Robot Epistemology

Author in paper [6] also makes use of the Wienerian idea in order to understand the various technological media as powerful ways to enhance power to intervene in the world and in nature, that is, the field of vision and perception that mimics the human gaze and understanding of the surrounding environment.

Symbolically, robots can be understood as a means of extending the human being, in their intervening ingenuity, refined and delicate manipulativeness, as well as in their power of increased information, communication and social interaction.

In the 20th century, during the beginning of automatization processes, many robots had weird and horrible shapes of mechanical devices, without natural likelihood regarding current living life, constituting apparatus that fulfilled several specialized functions in restricted environments very distant from quotidian life.

They had a great role in specific activities related to automobile industry line of production, scientific laboratory experiments, situations of demanded great precision and risk to human life, unmanned rescue operations, probes exploratory, surgical artifacts and other specialties of telerobotics.

Much of the services performed by robots used to be controlled step-by-step remotely, through continuous or sporadic human interventions. In a few cases, there was greater autonomy given to the machine to make decisions based on programming of predictable and unpredictable situations, without human action.

To some extent, the analogical approach in levels of human-robot similitudes aims to transform the interaction between man and machine as something more harmonious, cohesive, friendly, pleasant and affectionate, in view of the quality of the relationship between humans and their fellows of the same species.

In this sense, one has to think about how the accumulation of sensory robotic experience could effectively expand the perceptual field of vision. Would there be a possibility of turning the robotic gaze into an intelligent and sharp gaze, which not only sees visual objects, but would have the ability to give meaning to what it sees? In this case, what would the vision of beauty be interpreted by robotic sense? These are some valid questions in the context of this chapter. To enter the territory of such subtlety, it would be necessary to enhance the recognition of values such as empathy, for example, but the ability to demonstrate empathy in the most natural way possible, without sounding artificial and misleading. This can only be achieved if the human-machine interaction manages to constitute itself in a friendly interface that provides comfort, company and pleasantness.

The philosophical and epistemological bases of such questions can certainly illuminate the next steps in the development of robots that are more humanized or identical to what we are trying to express from now on as a sense of humanity: empathy, tenderness, affection, solidarity.

During the covid-19 pandemic period, many scientific studies demonstrated the difficulty of recognizing faces hidden by face shields, requiring attention to be focused on other relevant identification details, such as: the eyes, the way of looking, the hair, the physical body and the characteristic gestures of each person. But how could sophisticated robots recognize such details and identify the individuals they would have to interact with to carry out their activities, in the case of social robots?

In this regard, ethical precepts in the robotic field become more than necessary, as we are entering an era in which the vision and subtle perception of artificial intelligence will need to be increasingly employed to also identify, recognize and read human needs promptly. How to understand the symbology of a human gesture correctly, without offering a threat to the detriment of misunderstanding the reading of the gesture?

However, human facial recognition also presupposes a mode of social interaction, creation of affective bonds, feelings of empathy, belonging, and sense of friendship. Would it be possible for the new generation of robots to be able to differentiate and create significant associations based on facial recognition and the set of gestures of the face and the human body is a question of concern.

For author in paper [9], the period of the pandemic showed the growing importance of social robots with new facilitating functions: the so-called communication robots, and integrative social robots, also called responsible robots, which participate in sociability contexts, communicating themselves, and interacting at a distance, thus creating new forms of telepresence (telerobotics).

In this context, machines in general are seen more in their capacity as partners of social interaction. Of course, this leads us to revisit our understanding of the notion of sociality, which is changing with the development of artificial intelligence.

From this visionary field, robots tend to imitate more and more the physical forms, behavior, feelings, affection and sociability of human beings, including the character of an ethics of respect and care for human beings (roboethics).

Surely, this will imply in the perception and vision from such robots/androids or electronic persons regarding the human species in the future, and during the next generations of robotics.

Undoubtedly, post-pandemic social habits raise philosophical and epistemological questions of great interest for the development of robotic technology for the next generations, in the field of meticulous vision and sharp perception of identity.

1.5 Disruptive Challenges in the New Age Robotics

From a philosophical and epistemological point of view, will it be that with the current disruptive development of social robotics, we would not be entering a next stage of technology development highly interactive, participatory and completely autonomous in its decision-making and decision-making strategies of high complexity?

Today's rising robotics are enabling not only the future existence of robots that can almost fully perform many everyday human tasks, but also interacting with us from the perspective of a being similar to us, with identical or nearly identical features.

The humanoid appearance of future robots looks like it will no longer be a work of science fiction, but rather a future that could soon happen.

This, to a certain extent, should confuse our visual perception between what is virtually human in identity appearance, and what is actually human in reality, as everything seems to converge towards unavoidable approaches that facilitate technological friendliness.

According to [8], the uniqueness in the technological field will mean a radical paradigmatic disruption in terms of the machines' effective autonomy to act and decide on their own, naturally according to their own embedded and emerging logic of reasoning.

In this context, what would the field of vision of emerging and unique robots look like? How will we come to be seen among them as beings identical and distinct in nature? Partners, friends, enemies or competitors? How will we be affectively seen regarding robots´ vision and perception? These ae also questions of concern.

In view of this, perhaps, in the near future, robotic ethics will become an equalizing principle in defining the parameters of conduct for emerging robots in order to signal the defense and protection of human dignity at all costs.

Such algorithmic procedures will need, at a given moment, to turn into an effectively implementable, reconfigurable, and continuously improveable reality, to account for new realities of demanding behavioral conduct and correction of undesirable dysfunctions.

At some point in the future, the robotic singularity may develop the ability of robots themselves to design, build, repair, and reinvent themselves.

The robotization of society in various integrative, interactive, participatory and even humanizing molds and processes will be a challenge to be faced with great expectations, concerns, optimism, pessimism, and frightening fears.

Quoting ideas of author in paper [2]: "We become what we behold, we shape our tools, and thereafter, our tools shape us", referring to Mcluhan´s media morphogenesis dynamics [6], nothing is more provocative than pointing out that robots will develop their tools that will completely reinvent and redesign them.

In this way, as singularity becomes a concrete reality experienced in the daily life of technological society in the future, we will have to go through a process of complete robotization of social processes, interactions and relationships, being continuously immersed in fully complex robotized environments.

1.6 Conclusion

This work was an effort to try to interpret and understand some aspects related to the new generation of robotics, with the unimaginable advances that may come in the coming years, and which will lead to a great revolution in our way of social and affective relationships with the so-called robots social.

The entire robotization process of contemporary society should bring new philosophical and epistemological questions in the field of robotic knowledge, human and machine vision and perception, ethical and aesthetic values, and new relationships to be thought of with such entities materialized in the social field.

Such advances could bring a whole lot of anxieties and possibilities. The possibilities would need to be evaluated well and the anxieties would need to be courageously faced, so that excesses can be contained, and the ethical harmony in the relationship between robots and human beings can be properly preserved.

The technological and robotic singularity will symbolize a new era of robotics, with the rise of emerging actions and behaviors, which demanded a controlling functional ethical filter to give affective support to the different forms of social relations that can exist in the new technologized and robotized society.

In the dimension of the evolution of artificial intelligence, neural networks in line with robotics, the faster we can predict, diagnose and prevent misuse of future technologies, the better it will be.

Speculative and imponderable hypotheses may not be configured as concrete in the future, but they may give clues on how to face the imperceptible in continuous progress.

References

1. Assmann, Hugo. (2004). Curiosity and the pleasure of learning: The role of curiosity in creative learning. Editora Vozes, 2nd Edition, Rio de Janeiro, Brazil.
2. Culkin, J. (1967). A Schooman´s Guide To Marshall McLuhan. *Saturday Review, 51-53, 70-72.*
3. Freire, Paulo. (1996). Pedagogy of autonomy: Knowledge necessary for educational practice. Paz e Terra, São Paulo, Brazil
4. Lütkebohle, I., Peltason, J., Schillingmann, L., Wrede, B., Wachsmuth, S., (2009). The Curious Robot: Structuring Interactive Robot Learning. Applied Informatics Group. 2009 *IEEE International Conference on Robotics and Automation.* 4156-4162, doi: 10.1109/ROBOT.2009.5152521
5. Maturana, Humberto R. & Varela, Francisco J. (1972). *Autopoieseis and Cognition: The Realization of the Living.* In Series, Boston Studies in the Philosophy of Science, vol. 42, D. Reidel Publishing, Holland.
6. Mcluhan, Marshall. (1994). *Understanding Media: The Extension of Man*, The MIT Press, Reprint Edition, USA.
7. Merleau-Ponty. (2005). Phenomenology of Perception. Translated by Colin Smith. London, Routledge.
8. Kurzweil, Ray (2003). *The Singularity is Near: When Humans Transcend Biology.* New York, Penguin Group (USA).
9. Seibt, J. (2016). Integrative Social Robotics—A New Method Paradigm to Solve the Description Problem *and* the Regulation Problem? in: Nørskov, M./J. Seibt, *What Social Robots Can and Should Do—Proceedings of Robophilosophy 2016 / Transor 2016,* forthcoming.
10. Wiener, Norbert. (1985). *Cybernetics or Control and Communication in the Animal and Machine.* The MIT Press, Cambridge, Massachusetts, USA.

2

Advances in Robotics: Applications and Advantages

Manisha Vohra
Independent Researcher, India

Abstract: Robotics has progressed a lot. There have been a lot of advances in robotics. Now there is new age of robotics which has many possibilities. It can be used in various sectors. In this book chapter, with the advances in robotics and the presence of new age of robotics, the application of robotics in different sectors like healthcare and agriculture will be discussed. Along with it, the need of robotics in these both sectors and advantages of application of robotics in these sectors will also be discussed in the chapter.

Keywords: Robotics, healthcare, agriculture, **rehabilitation robot**, agricultural robot, etc.

2.1 Introduction

Robotics has witnessed tremendous growth in the recent times. There have so many advances in robotics. In today's time, there is a complete new age of robotics. Karel Čapek who was a Czech writer had made use of the word robot in a play in the year 1920. This word was derived from the word robota which in Slavic languages means labour. After the word robot was used in the play, it gained popularity. Robotics can be used in various sectors. In this chapter, we will be discussing how robotics can be useful for application in healthcare and agriculture sectors.

Healthcare sector is a crucial sector. Development and advances in this sector are always worked upon and given importance. Any advances that can benefit this sector in a proven manner will be advantageous for this sector. Robotics in the recent times has gone through many advances. With the new age of robotics now there are different kind of possible applications of robotics in healthcare. Especially in a period like pandemic where help of technology is needed to be explored, exploring regarding the application of robotics in healthcare sector can be useful for this sector. In fact, not just in healthcare sector but also in the agriculture sector, robotics can be useful. We all know agriculture sector is the backbone of our society. This sector is majorly responsible for taking care and fulfilling the food requirements of the people. It provides with variety of food produces like vegetables, fruits, etc. Hence it is a sector with extreme great importance. Robotics which is having various capabilities can prove to be of great use in agriculture sector as well. The different applications of robotics in healthcare and agriculture sector will be discussed further on in the chapter.

2.2 Need for Robotics in Healthcare Sector

Any kind of technology which can help this sector in a proven way will be beneficial for this sector and also the ones associated with this sector. Healthcare being an extremely important and crucial sector, it requires cutting edge technology which can help in a proven way. In today's world, where lives are so fast paced and occurrence of events like pandemic are witnessed which have huge impact on people's health, becomes more important and necessary to different technology based resources which can be of help in different ways in the healthcare sector. Hence there is a need for such technology in healthcare sector. Robotics has the required abilities that can help this sector and people related to it and benefit this sector. The advantages robotics can provide to healthcare sector are amazing which will be stated further on in the chapter.

2.3 Applications of Robotics in Healthcare Sector

A literature review was carried out to view and understand different possible applications of robotics in healthcare sector. It was noticed that in healthcare sector, there can be numerous possible applications of robotics. Some of the applications from the literature review are discussed as follows.

Big advances in minimal invasive surgical care delivery is robot-assisted surgery. The authors in [2], carried out a study in which analysis of operational policies is done that have potential to be implemented at hospitals in order to realize clinical outcome benefits and along with it at the same time to also control health care delivery cost. This is enabled by technologies that are advanced, like for example, a surgical robot. This particular study's focus is a robot which is a surgical one and it is a da Vinci surgical system. Hysterectomy is the procedure of the surgery which is context of the study. The authors showcase an integrated methodological approach's application which helps to find out and also to do analysis of policies that are actionable and which can be implemented by a hospital. Few vital contributions of this paper include showcasing of different things like hospital-level policies. These policies can be helpful in realizing clinical result and cost benefits of a surgical robot. Another vital contribution is that the patient condition's criticality is an important determinant of surgeon learning in robot-assisted surgery [2] and likewise there are more such different key contributions of this article. This study by the authors provides useful outcomes.

Artificial intelligence (AI) can also be helpful in healthcare. Outside the operating room, AI and robotics technology can be implemented for various tasks such as admission of patient in digital mode, monitoring of vital signs of patients, delivery of food and medicines [10]. Paper [5] explains medic robot's utilization for various work like providing delivery of food the patients, providing delivery medicines to the patients, etc. [5].

The authors in paper [4], have given the description of the telepresence robot system which is specially designed here in this case by the authors in order to improve the elderly people's well-being by providing them support to perform without any dependence the usual daily activities, provide facilitation of social interaction so as to come out of a sense of isolation socially and loneliness and also as well as to also provide support to the professional caregivers in everyday care. To find out regarding the acceptance of the robot system which was developed, an evaluation study was carried out. This study included two different potential user groups i.e., of elderly people and professional caregivers. The developed system provides with the facility of video contacting with doctors, nurses, family and friends and it is a two-way contacting facility. This helps in strengthening relationships and social interaction. With the help of the evaluation study, the authors wanted to find the acceptance of the users with respect to the developed system and as well as wanted to know how both the user groups would in general perceive this system. Also, the authors were seeking to find out what are the user groups opinions, concerns and needs in regards to this developed system. A questionnaire was prepared for the evaluation study purpose. It was demonstrated from the results i.e. robot acceptance questionnaire, that the developed telepresence robot system's provided core functionalities are accepted by the potential users. The finding from this result corresponds with the data of the interview, in which users have given opinions they have about the robot system which were quite positive. Another point noticed was, the users who participated in this study, identified way more benefits as compared to concerns regarding utilizing this system [4].

In this paper [3], the authors for a foot-plate based sitting-type lower limb rehabilitation robot have proposed a planar hybrid manipulator system which is integrated with a passive serial orthosis. There is a modularity in the proposed robotic system which can be with ease upgraded to a standing-type body weight support (BWS) mechanism. Intensive limb training is required in order to recover effectively from motor impairments of lower limb and for this purpose the proposed robotic device is designed for sitting type gait therapy. The rehabilitation robot which is proposed in this paper is having different merits and it has been fabricated in-house and its functions are showcased for one leg in real-time, the reason being that the other leg eventually have synonymous dynamic and kinematic performance eventually [3].

2.4 Advantages of Robotics in Healthcare Sector

Based on the above discussed literature review based applications of robotics in healthcare, some of the prominent advantages of it are as follows:
- ➢ Robotics could help in delivery of food and medicines to the patients
- ➢ Robot system could be useful for helping elderly people to perform regular activities of day to day life independently and could also facilitate social interaction as discussed and seen in paper [4].
- ➢ Robot system could be utilized for rehabilitation purposes as discussed and seen in paper [3].

2.5 Need for Robotics in Agriculture Sector

Agriculture is not just an important sector but also one which is extremely necessary and useful as it produces food. In this sector also different technology which can help the agriculture sector in a proven way will be advantageous for this sector and also for others who are associated with this sector. Robotics can help in agriculture sector.

2.6 Applications of Robotics in Agriculture Sector

Here, in this case also to view and understand different possible applications of robotics in agriculture sector, a literature review was carried out. It was observed robotics can be utilized in agriculture sector as well, just like in the case of healthcare sector. Again, here we discuss below some of the applications from the literature review.

The authors in [11] developed a modular and precision terrestrial sprayer robot, i.e., the Precision Robotic Sprayer. This helps in precision spraying. The environment is benefitted with precision agriculture as it helps in decreasing the number of pesticides used in agricultural fields and it also decreases the amount of resources that are spent on machinery. Utilizing image histograms (vegetation index, local binary pattern (LBP), hue and average), leaf density is calculated based on a support vector machine (SVM) classifier and this calculation is done by crop perception system which the sprayer comprises of. The perception system was developed and tested. The results of the leaf density classifier showcased a high accuracy. The tests carried out prove that this work has the potential to increase the spraying accuracy and precision [11].

A modular agricultural robot was developed by authors in [8] for selective spraying of grapevines for the purpose of controlling the disease. Results point out that automatically the robot could detect and spray from 85% to 100% of the diseased area within the canopy. The robot was also able to bring reduction in usage of pesticide from 65% to 85% when compared to the canopy's conventional homogeneous spraying [8].

At present grapevines and orchards and grapevines are overall sprayed. Airflow assistance is utilized by most of the tree crop sprayers and bush sprayers. This generates movements in canopy exposing both the leaves sides to the spray. There is formation of huge coherent vortices which further contributes to improved coverage of spray. For the purpose of disease foci's selective treatments, a new close-range spot-spraying method which is air-assisted is evaluated here in this paper [7] by authors and it is promising for reducing pesticides. Minor areas of disease require treatment when they in initial development stage for precision spraying in close range and for this purpose, a spraying end effector (SEEF) which is a close range precision application was designed and was intended for installation on robotic arms. Hence it was of lightweight construction. The SEEF effector's evaluation was done by the authors. They evaluated it for a close range precision spraying process in vineyards. The results had advantage which was good but however there was a drawback as well which needs to be worked upon and overcome in the future.

Only the plants which require treatment, precision spraying enables its treatment that too with the right number of products. The authors in paper [1], have developed a solution which is based on a reconfigurable vehicle with a high degree of automation for the purpose of distribution of plant protection products in greenhouses and vineyards. Interaction among the spraying management system and autonomous vehicle is a solution which is having high content of technology. It enables allows safe as well as accurate autonomous spraying operations. For the purpose of evaluating system performance, in two different scenarios trails were performed. The two different scenarios were vineyards and greenhouses. As per the experimental results demonstrated by in the paper, good performance was showcased by the proposed architecture that too even in a scenario of a greenhouse which could introduce Global Navigation Satellite System (GNSS) signal attenuations and create multipath reflections. Without any significant degradation, all the trials were performed. Keeping an enough safe distance from the plants, in an autonomous manner the robot conducted out the path which was assigned between the rows for nearly about 8 hours. The smart spraying system took care of the product to be sprayed for the treatment. One thing can be stated from the preliminary tests conducted by the authors in the paper [1], that the platform showcases nice prospective to be able to conduct the work even in hard conditions. [1].

2.7 Advantages of Robotics in Agriculture Sector

Based on the above discussed literature review based applications of robotics in healthcare, some of the prominent advantages of it are as follows:

- ➤ Thanks to robotics application in agriculture, a precision terrestrial sprayer robot can be made which can be utilized for precision spraying as discussed and seen in paper [11].
- ➤ Agricultural robot can be developed and utilized for successful selective spraying of grapevines for disease control which even reduces the use of pesticide as seen and discussed in paper [8].

2.8 Conclusion

Robotics has grown over the years. In recent times, the growth and advances seen in it have made it possible to have innovations and utilize it in different sectors. The applications now possible with new age of robotics have opened doors to various new opportunities. In this book chapter, application of robotics in different sectors like healthcare and agriculture were discussed. Along with it, the need of robotics in healthcare and agriculture sector was also discussed and different advantages of the applications were stated in the chapter.

References

1. Cantelli, L., Bonaccorso, F., Longo, D., Melita, C. D., Schillaci, G., Muscato, G. (2019) A Small Versatile Electrical Robot for Autonomous Spraying in Agriculture. *AgriEngineering*, 1, 391-402.
2. Mukherjee, U. K, Sinha, K. K. (2020) Robot-assisted surgical care delivery at a hospital: Policies for maximizing clinical outcome benefits and minimizing costs. *Journal of Operations Management*, 66(1-2), 227– 256.
3. Mohanta, J. K., Mohan, S., Deepasundar, P., Kiruba-Shankar, R. (2018) Development and control of a new sitting-type lower limb rehabilitation robot. *Computers and Electrical Engineering*, 67, 330–347.
4. Koceski, S., Koceska, N. (2016) Evaluation of an assistive telepresence robot for elderly healthcare. *Journal of Medical Systems*, 40(5), 121.
5. Sahu, B., Das, P. K., Kabat, M. R., Kumar, R. (2022) Prevention of Covid-19 affected patient using multi robot cooperation and Q-learning approach: a solution. *Quality & Quantity* 56(2), 793–821.

6. Christoforou, E. G., Avgousti, S., Ramdani, N., Novales, C., Panayides, A. S. (2020) The Upcoming Role for Nursing and Assistive Robotics: Opportunities and Challenges Ahead. *Frontiers in Digital Health*, 2, 585656.
7. Malneršič, A., Dular, M., Širok, B., Oberti, R., Hočeva M. Close-range air-assisted precision spot-spraying for robotic applications: Aerodynamics and spray coverage analysis. *Biosystems Engineering*, 146, 216-226.
8. Oberti, R., Marchi, M., Tirelli, P., Calcante, A., Iriti, M., Tona, E., Hočevar, M., Baur, J., Pfaff, J., Schütz, C., Ulbrich, H. (2016) Selective spraying of grapevines for disease control using a modular agricultural robot. *Biosystems Engineering*, 146, 203-215.
9. Anghel, I., Cioara, T., Moldovan, D., Antal, C., Pop, C.D, Salomie, I., Pop, C.B., Chifu, V. R. (2020) Smart Environments and Social Robots for Age-Friendly Integrated Care Services. *International Journal of Environmental Research and Public Health*, 17(11), 3801.
10. Zemmar, A., Lozano, A. M., Nelson, B. J. (2020) The rise of robots in surgical environments during COVID-19. *Nature Machine Intelligence*, 2, 566–572.
11. Baltazar, A. R., Santos, F. N. d., Moreira, A.P., Valente, A., Cunha, J. B. (2021) Smarter Robotic Sprayer System for Precision Agriculture. *Electronics*, 10(17), 2061.
12. Pou-Prom, C., Raimondo, S., Rudzicz, F. A (2020) Conversational Robot for Older Adults with Alzheimer's Disease. *ACM Transactions on Human-Robot Interaction*, 9(3), Article 21.

3

Prosthetic Arm for Rehabilitation Robotics

Aishwarya S, S. Srinivasan, Sushmitha M. and Veena Hegde
Department of Electronics and Instrumentation Engineering,
B.M.S. College of Engineering, Bangalore, Karnataka, India.

Abstract: Electromyography (EMG) signals can provide benefit to biomedical/clinical applications, etc. Muscle EMG signals through surface electrodes or needle electrodes, require advanced methods. In our work, we implement a hardware set up of the signal acquisition system and develop code to realise the servo actuation in real time to observe the muscle activation and the movement of the servos accordingly. We further analyse the acquired data in the time and the frequency domain in order to identify most prominent features which can be used to classify the various gestures of the wrist and the finger movement by using various signal processing techniques. These features are compared and verified by using data from existing databases online such as the Ninapro. A performance comparison study of various EMG signal analysis methods is also provided. This will help to further develop a powerful model for its application in the prosthetic arm. A 3D model has also been in the making that would house the hardware in future iterations of the project.

3.1 Introduction

There are various aspects to be considered before designing a prosthetic arm. The literature survey being a crucial part enlightened us on the existence of various online open-source databases where we could learn about the nature of EMG signals and how they occur, how they can be captured and observed. The Ninapro database, a freely accessible resource aimed at assisting advanced research is helpful. The database is compiled by a group of people working together. Surface electromyography signals from the forearm were recorded. While volunteers perform a preset task, kinematics of the hand and wrist a collection of actions and postures. The database comprises information from 27 healthy participants. 52 finger, hand, and wrist movements are performed. Both intact and amputated subjects' data are collected. It offers a benchmark classification result using a range of feature representations and classifiers, in addition to explaining the acquisition strategy and processing procedures. The public surface electromyography (sEMG) database is the most up-to-date database available. Taken into consideration in terms of electrode placement, device calibration, and it's all about data capture and synchronization. Short videos are used to instruct the stimuli, which the subjects are requested to imitate. We can follow a similar procedure for data acquisition from our own subjects. For the subjects, this makes the protocol exceedingly straightforward, stress-free, and fatigue-free. In a continuous prediction setting, a benchmark examination of a number of common feature extraction and classification approaches finds that the best performing algorithms attain an accuracy of roughly 76 percent [1].

We cross verify the accuracy results of classification on our MATLAB 2019 Ra. This result also shows that, when combined with a suitable classifier, very simple features like MAV can perform equally as well as more complicated mDWT (Modified Discrete Wavelet Transform) or short-time Fourier transform (STFT) features. The accuracy of categorization drops dramatically when the subject's body mass index (BMI) rises, according to a multiple regression analysis of classification accuracy with regard to many subject parameters.

EMG signals consist of the activity of distinct motor units superimposed on one other and there are different approaches for breaking down an EMG signal into its constituent parts [2].

This helped us gain a lot of knowledge about the detailed behavior of EMG signals and how they can be decomposed.

We got an idea of using wavelet transform as a feature extraction technique. Using a mother wavelet matrix, a method for extracting relevant features for forearm electromyographic (EMG) signals was proposed (MWM). The combination of several mother wavelet functions improved the EMG signal analysis after extensive research on 324 mother wavelet functions. We try to apply similar signal analysis technique for feature extraction. The ideal sensors for feature extraction were chosen in terms of surface electrode matrix (SEM) and needle electrode matrix (NEM) from among many placed electrodes on the patients' forearms in this paper. This research looked into six different statistical feature vectors. EMG signals are essential in a variety of sectors. In this paper, a high resolution, eight channel system has been designed to suit the requirements of EMG data gathering systems. It is cost effective [4].

This gave us an understanding of designing our own data acquisition system. A low-cost system for capturing EMG signals and their utilization provides flexibility for users for visualizing the signal during activity and for exporting the data in desired formats for signal processing. In contrast to other acquisition systems based on instrumentation amplifiers, this system consumes less power, as it uses only one amplifier IC for 8 channels, as opposed to the usual three amplifiers for instruments with amplifiers. It can be extended from 8 channels to 16 channels [5].

The device that is utilized to help the program's activities should increase electrical activity in the motor unit while reducing the user's mental strain. Electromyography (EMG) is a technique for determining whether or not electrical activity exists in the musculoskeletal system. As a result, the rehabilitation device should assess typical people's surface EMG signals and incorporate them into the gadget [6].

We learnt how we can recognize electrical activity and map that to hardware. The signal is acquired using a surface electromyography method for non-invasive muscle evaluation (SENIAM). Different time-domain characteristics were extracted. It is critical to obtain EMG data. The precise features can be used for real-time rehabilitation control. The goal of this research was to examine features in the temporal domain as soon as possible so that they might be used in future efforts such as pattern recognition or classifiers which is what we also aim to achieve. The importance of this study is to identify the percentage error of the features in order to acquire the optimum feature performance. In the time domain, the MAV exhibits the best performance according to this paper. The tiny error percent and the features that help to create the rehabilitation system device with higher device control system performance are features that help to design the rehabilitation system device with better device control system performance [6].

Manual lifting is considered one of the vital methods of material handling. Improper lifting techniques can lead to musculoskeletal disorders (MSDs), with overexertion being the most common cause. To avoid overexertion, electromyography (EMG) signals are utilized to monitor the workers' muscle health and to determine the maximum lifting weight, lifting height, and number of repetitions that they can perform before becoming fatigued. This paper taught us about how muscle fatigue affects the muscle activity and that more MUAPs are needed to lift [8]. Several EMG processing algorithms and distinct EMG features that indicate fatigue indices in the time, frequency, and time-frequency domain have been introduced by previous studies [8].

Our motivation was also to work on finger movement control. During a set of typing exercises, myoelectric data was gathered from the forearm muscles of twelve normally limbed participants. These data were used to evaluate a variety of categorization systems, each with a unique set of system element options.

3.2 Design methodology

We first identify the points of interest for acquiring signals accurately. Motion artefacts, electrode misplacement, and noise interpolation are just a few of the elements that affect the EMG signal. Filtering, rectification, baseline drifting, and threshold levelling are used to acquire additional information from EMG signals. EMG activity only over the forearm and upper arm muscles was measured for feature extraction.

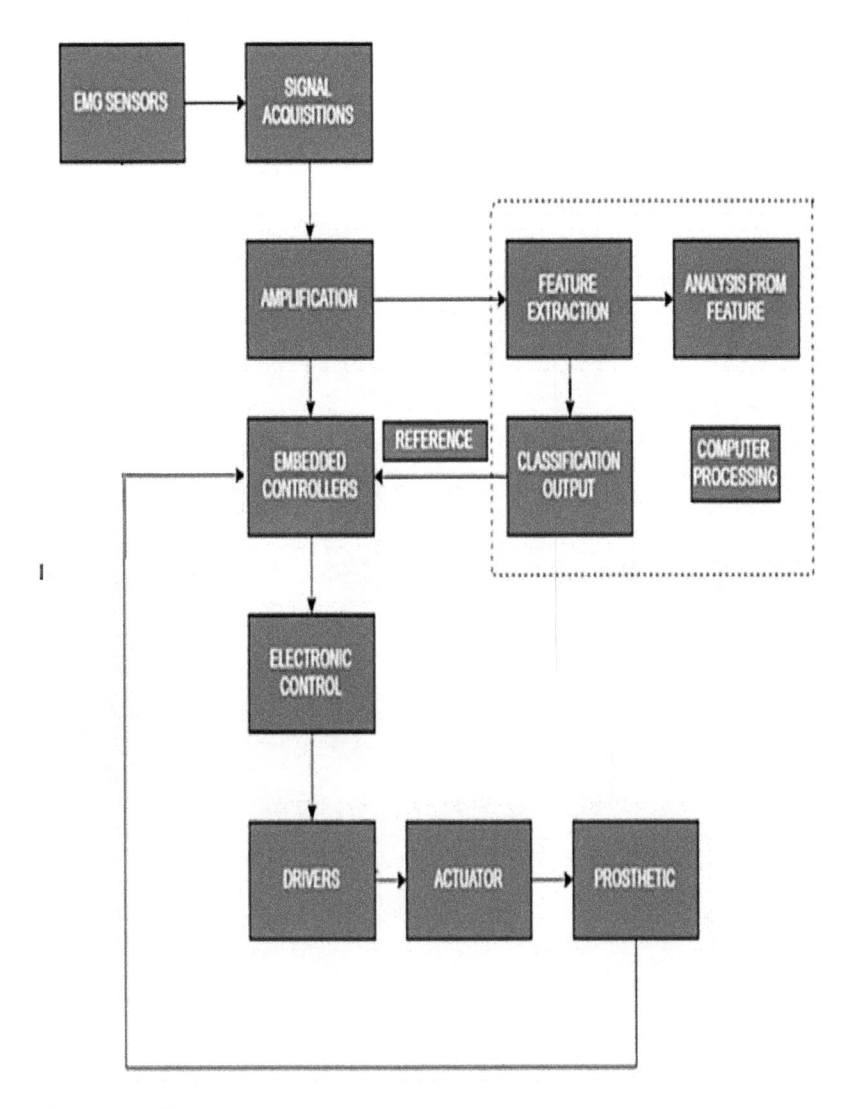

***Figure 3.1** Block Diagram*

The given block diagram is one that we developed which would give us at most accurate functioning of the entire prosthetic arm. Continuous testing and debugging of the system would give efficient results. Four forearm muscles were selected for their major role in stabilization and movement of the upper limb upon looking into resources like SEINAM and also by physiotherapy experts.

1.Flexor carpi radialis (FCR),

2.Extensor digitorum (ED),
3.Extensor carpi ulnaris (ECU),
4.Extensor carpi radialis brevis (ECRB)

3.2.1 Data Acquisition

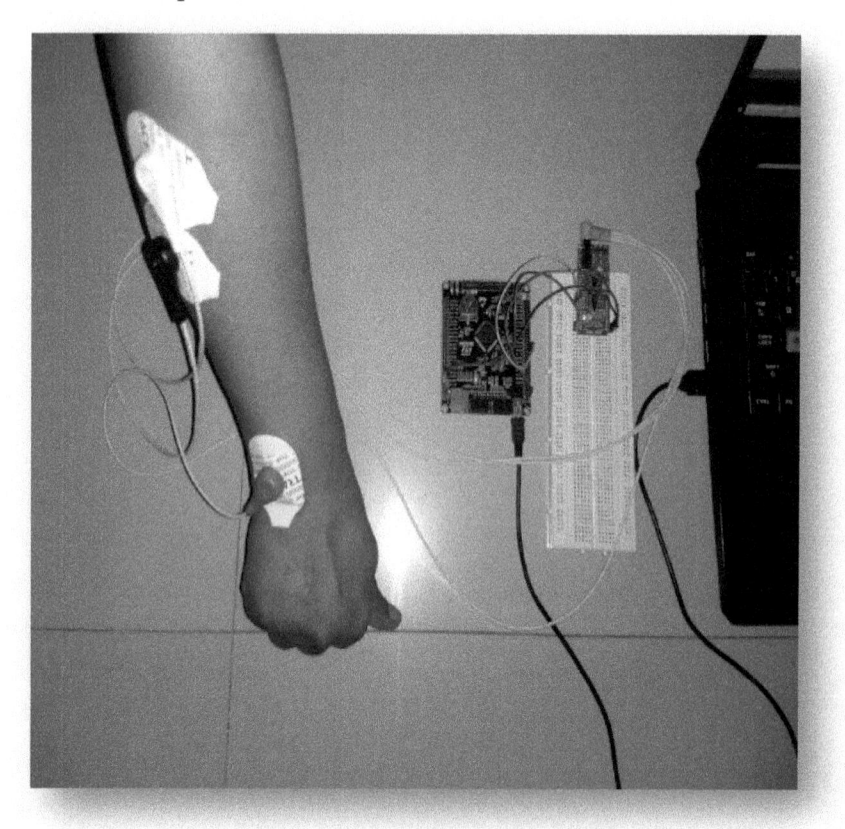

Figure 3.2 Data acquisition hardware setup

STM32 Microcontroller, which has an analog input ADC Channel is used. It Can be used for multichannel sampling as well. The STM32 Microcontroller was chosen as it has 16 channels of ADC and can work extremely well with multiple channels of the sensor board.

It has a DSP function and has a clock frequency of 168Mhz, which is good for data acquisition to avoid losses of information. The electrode cables are connected to a 3.5mm Audio jack and connected via the EMG Click Board on the breadboard. The STM32 is programmed via USB Cable in DFU mode.

The entire setup is now powered by connecting the USB Cable to the laptop and data from the sensors (electrodes) are acquired serially. The same principle can be used for multi-channel sampling using the DMA or the Polling Interrupt method to switch between different ADC channels for conversion.

Skin preparation is a crucial step as the electrodes are made of Ag/AgCl and are quite sensitive, and hence the hair and dirt present on the skin may cause interference with the signals acquired. According to the papers and SENIAM, the skin has to be prepared by removing the hair at the surface of interest point and applying some form of spirit or sanitizer to clean the skin surface.

The electrodes were placed according to the instruction, one in the middle of the muscle and the other at the tail end of the muscle as that is the place where the strength of the signal is the strongest. The LL Electrode was either placed on the elbow or wrist as the bony reference electrode. This reference electrode is needed so as to reduce noise and get a plain signal from the stationary part of the electrode for the differential amplifier present on the signal conditioning board.

An opening and closing of the fingers with 1 second interval for 15 seconds was made and the data was visualised on the serial plotter of the Arduino IDE. This data which is the ADC data ranging from 0-4095 for 12-bit data or 0 – 1023 for 10-bit data is collected from the serial monitor. These files are converted to CSV format and can be easily imported into the MATLAB or Jupyter notebooks environment for further signal processing and analysis.

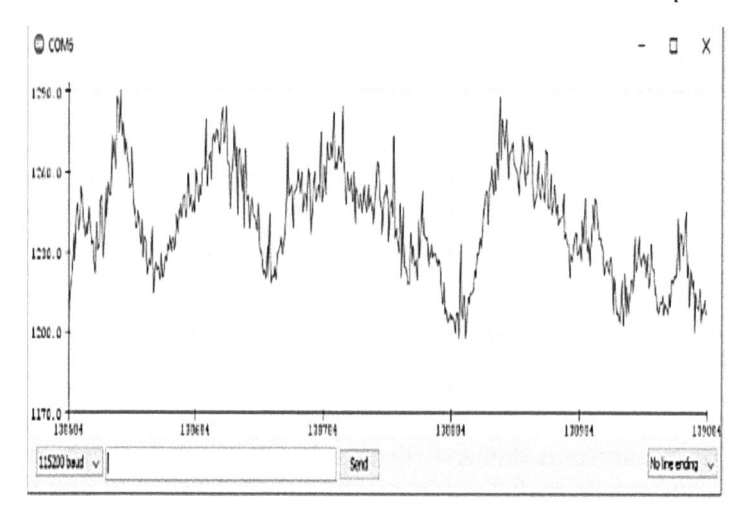

Figure 3.3 *Serial plotter on Arduino*

The above is the serial plotter output of the data that we acquire in real time and hence we can see that the upper spikes of the entire signal are when the extension action is taking place of the wrist as we have taken this signal when the electrodes are placed at the extensor carpi radial is brevis as seen in the pictures previously.

As seen in the above picture, the EMG signal has some noise and equipment artifacts, hence we can get a better resolution on the ADC by using an external voltage reference of 3.3v. So now our ADC values will range in the numbers for the voltage of 0-3.3v rather than 0-5v and hence we can see a cleaner signal with better variation as shown The output waveform we get for that correction is as shown below in the figure.

Robotics and Modern Computer Vision 31

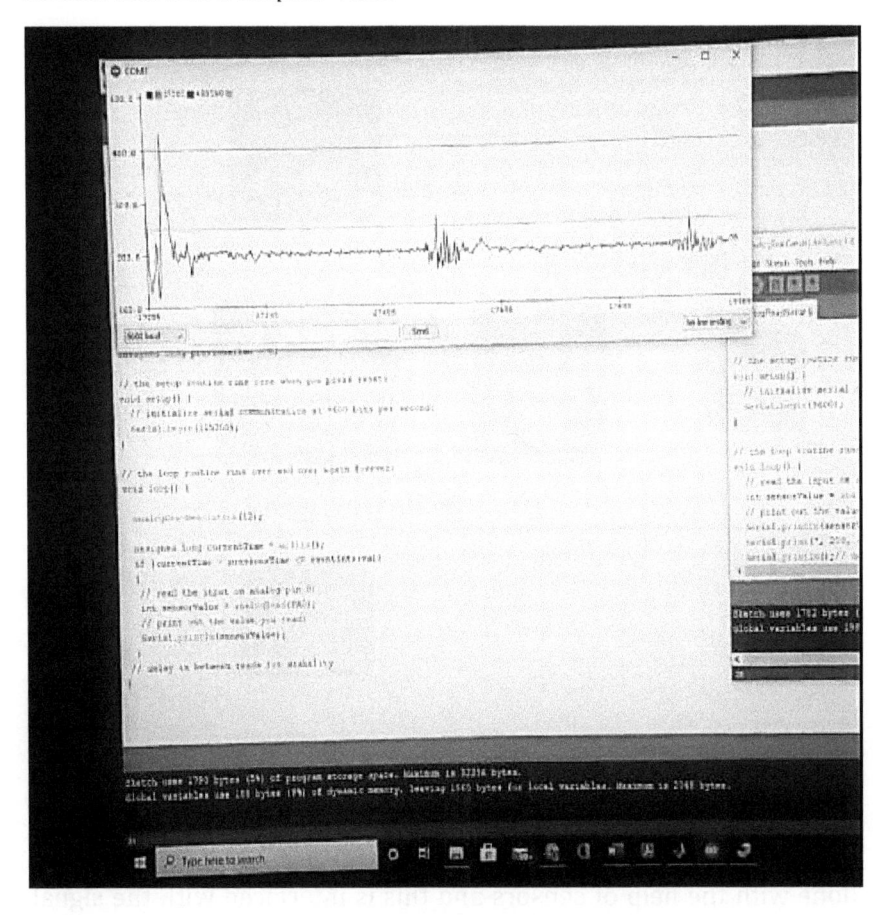

Figure 3.4 *Variation of signal with reduced noise*

It is observed in the figure that the peaks occur whenever there is a movement of the hand, in this case wrist flexion and extension has been performed. The difference between noise and the peaks can be observed clearly and would help us setting the thresholds for further detection of the signals. The range in the above figure is around 150 – 400 in terms of ADC raw values upon changing the range.

3.3 Hardware Architecture

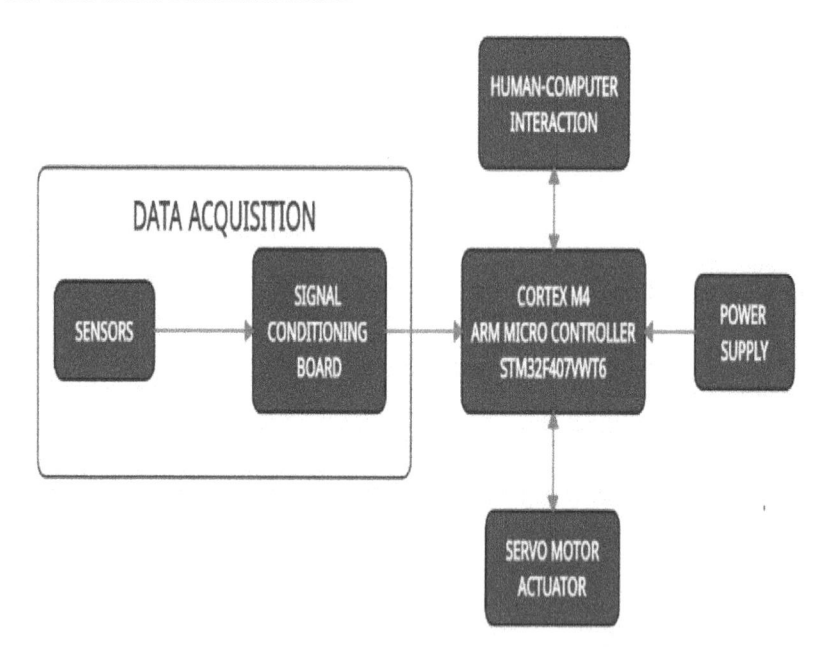

Figure 3.5 Hardware architecture

Our hardware architecture can be given in the form of a block diagram as shown above. It can be seen that the data acquisition is being done with the help of sensors and this is interfaced with the signal conditioning board which is further interfaced with the STM32 Microcontroller according to the connections specified earlier. The STM32 is interfaced with the PC and data is transferred accordingly. The system is externally powered by a 9v battery. After the thresholding is detected, the servo actuation is done accordingly.

Robotics and Modern Computer Vision

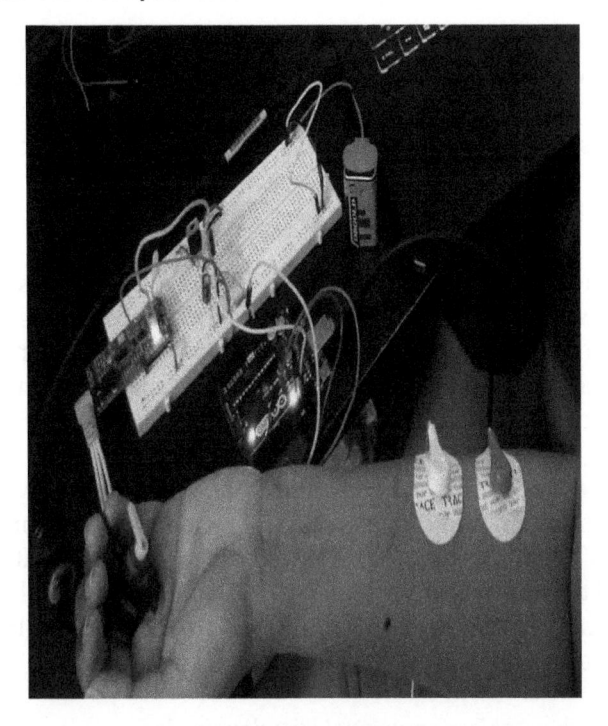

Figure 3.6 *Hardware setup for servo actuation*

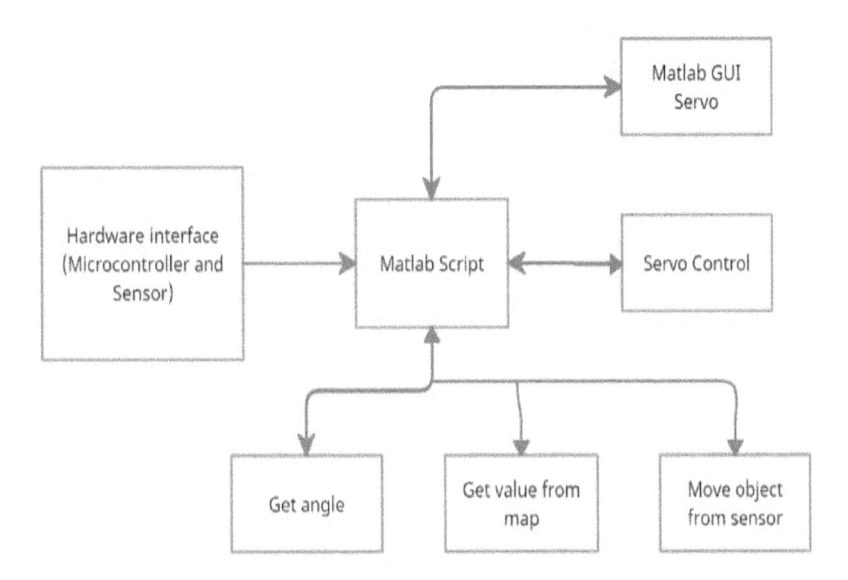

Figure 3.7 *Hardware implementation*

Our final model involves the digital twin in hardware as well as software where the signal obtained in real time is compared with the threshold set for servo actuation in the software GUI as well as the hardware servo motor. In order to verify the signal acquired and translate it to motion both in hardware and software, we developed a digital twin.

This GUI shows a servo in software that behaves exactly as a servo in hardware would and hence verifies the action. Thresholds of the signal values in voltage are mapped to servo angles in an array usually varies based on the strength of the signal and the duration for which the electrodes are being used for.

When these thresholds are reached, then the servo moves to that particular position and displays the current servo position and that behaves as a feedback loop and the servo adjusts its position based on the real time values of the signal. The same action translates to the hardware servo. The lookup table shows the servo angles for the range of the signal seen on the MATLAB environment.

Table 3.1 Lookup table of servo angle for signal value

Threshold	Servo angle
<1.8	0
1.8-2	0
2-2.0	30
2.0-2.2	60
2.2-2.4	75
2.4-2.6	90
>2.6	90

Robotics and Modern Computer Vision

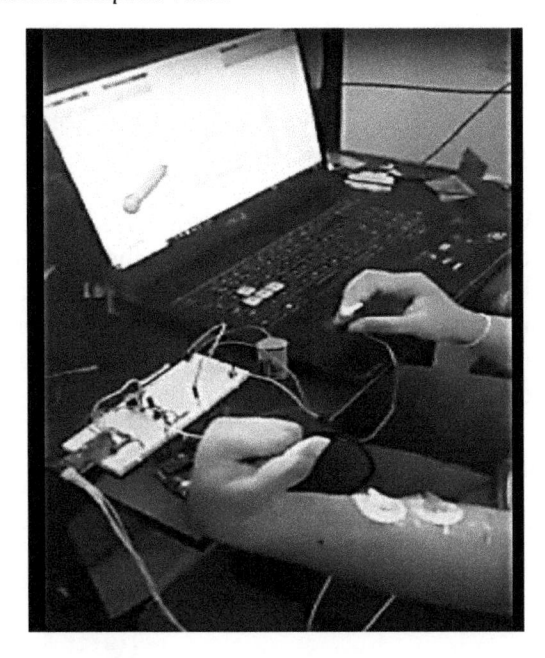

Figure 3.8 *Real time movement of servo in the GUI as well as hardware according to hand movement.*

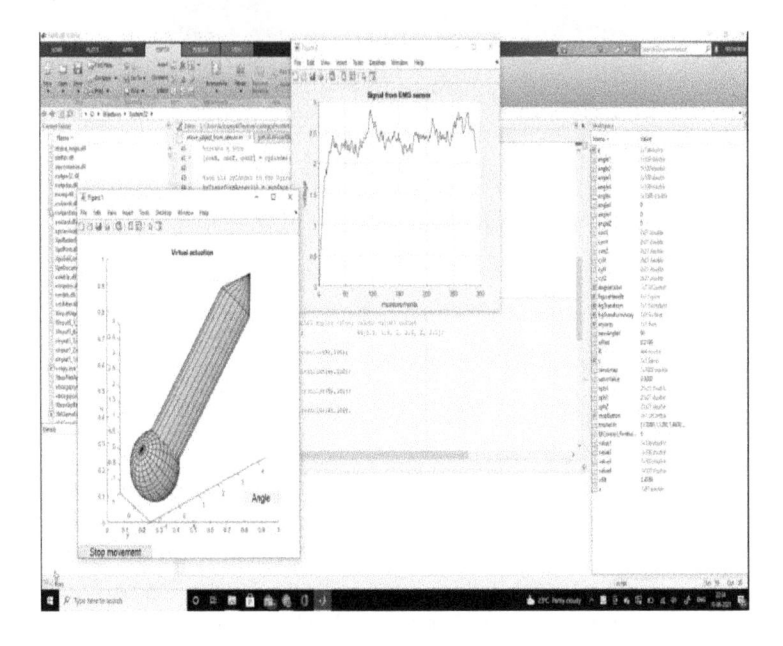

Figure 3.9 *GUI and the real time signal from the sensor showing servo actuation.*

The software GUI is programmed to the thresholds in such a way so as to simulate the range of the wrist from 0-90 degrees which is also translated to the hardware servo which is connected to the microcontroller on the digital pin.

The 1.5ms PWM signal is the range of 90 degrees of the servo movement and the intermediate angles are reached according to the lookup table based on the infinite for loop which acts as a PID controller with only the P coefficients which adjusts the servo values as feedback when the threshold value is lesser than 40 degrees as this requires more precise actuation.

3.4 Data Analysis

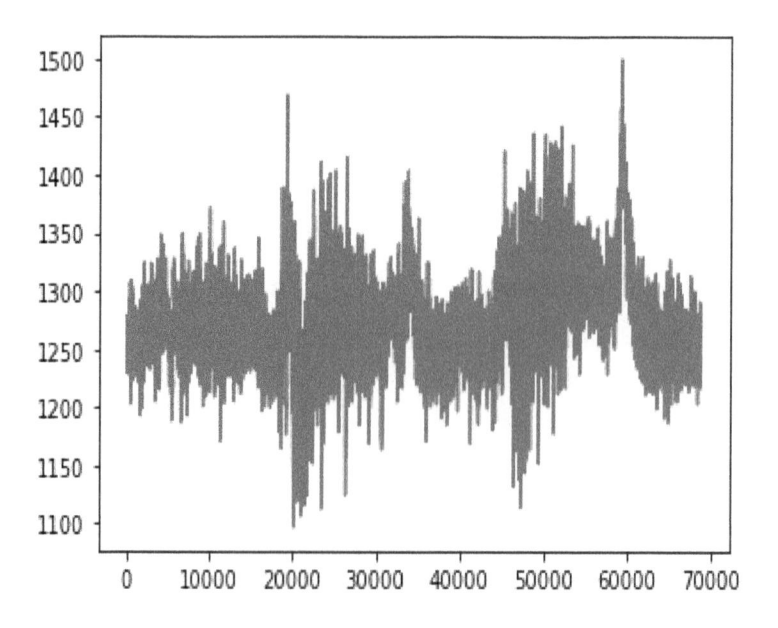

Figure 3.10 Raw sEMG signal - Extensor Digitorum

Raw sEMG data – The surface EMG electrodes are used to acquire the EMG from desired muscle when performing a task/gesture.

Unit conversion of data - The acquired data has to be converted to the corresponding voltage value to study the nature of the signal. We see that the EMG data collected lies in the typical EMG signal range of 2-10mV. Conversion of units is done using the following formula:

$$EMG_{mv} = \frac{\left(\frac{ADC}{2^n} - \frac{1}{2}\right) VCC}{G_{EMG}}$$

EMG$_{mV}$ - EMG value in mV
ADC - Digital value sampled from the channel
n - Number of bits of the channel (10)
VCC - 5000 mV
G$_{EMG}$ (Gain) – 1000

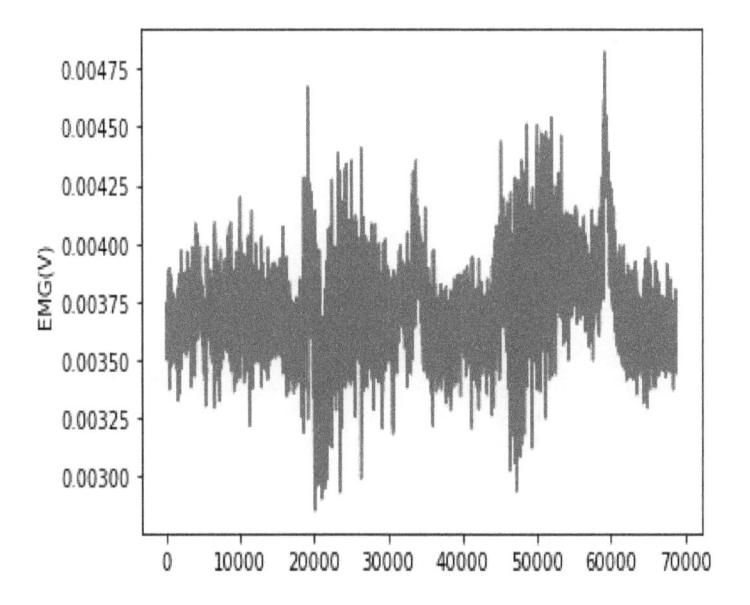

Figure 3.11 EMG data converted to volts

Fast Fourier Transform - Muscle tiredness is typically assessed using frequency domain analysis of changes in electromyographic (EMG) signals over time. Although Fourier analysis implies signal stationarity, which is improbable during dynamic contractions, it is commonly utilised in these assessments. For a time domain signal x(t), the forward and inverse transforms of a continuous time Fourier transform (CTFT) are defined as

$$X(j\omega) = \int_{-\infty}^{\infty} x(t) e^{-j\omega t} dt$$
$$x(t) = 12\pi \int_{-\infty}^{\infty} X(j\omega) e^{j\omega t} d\omega$$

When applied to EMG data, the Fourier transform reveals how the signal's power is spread across different frequencies, allowing the signal to be analysed in the frequency domain. Any periodic signal can be represented as a sum of sine and cosine waves of variable amplitudes, frequencies, and phases using the Fourier Transform. Each sinusoidal component's amplitude (A) or power can be analysed as a function of frequency, yielding the amplitude (A).

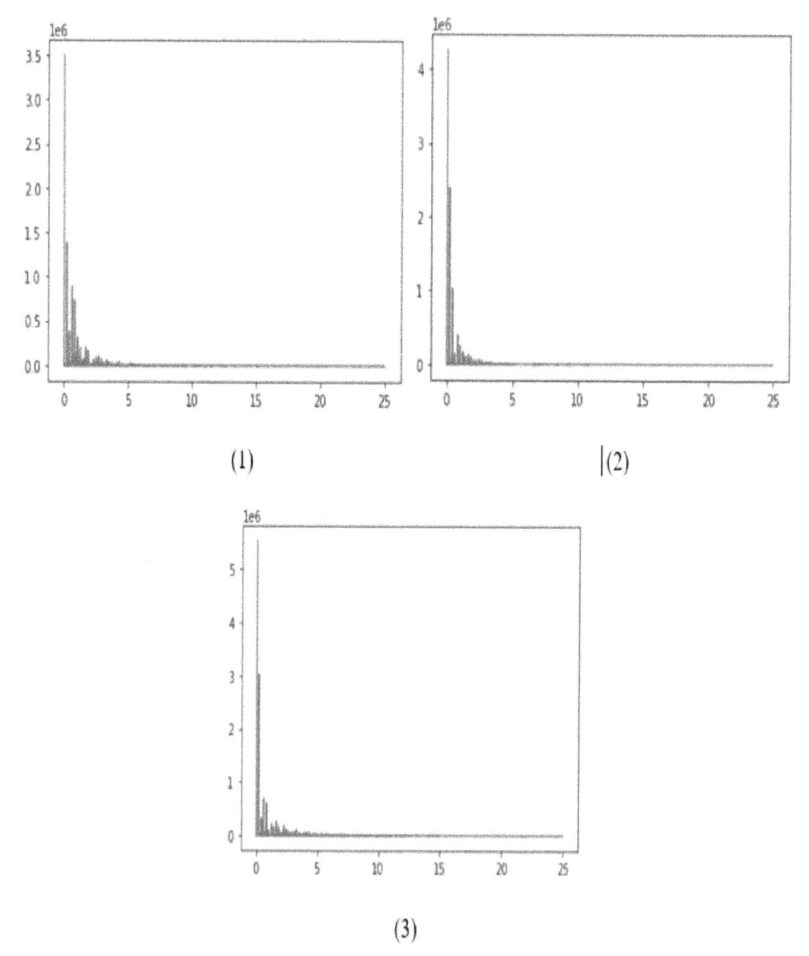

Figure 3.12 *Fast Fourier Transform of sEMG signal collected from (1) Extensor Digitorum, (2) Flexor Carpi Radialis, (3) Extensor Carpi Ulnaris.*

It is observed from the FFT graph that the prominent frequencies are seen till a little more than 500 Hz and after that, the signal is made up of equipment artifact noise. Hence, we need to design a low pass filter with cut off frequency as 500Hz.

Baseline noise removal - Baseline oscillations and low frequency noise can degrade the quality of an EMG signal, disrupting the motor unit action potential (MUAP) extraction, analysis, and classification process. Here, the baseline noise is removed using three techniques; removing the mean value of the signal, filtering using a low pass Butter worth filter, and rectifying the filtered signal.

Mean value removal – The removal of the signal's mean value is a crucial step in the EMG signal processing. This aids in normalising the signal, which is necessary for improved classification accuracy because it involves comparisons with a variety of other signals. The mean adjusted signal's average is zero. Because of the intrinsic variability of EMG signals, clinical interpretation of surface EMG signals is difficult.

$$EMG_{Mean-corrrect} = EMG - \left[\frac{\sum EMG}{L_{EMG}}\right]$$

Mean corrected signal is obtained by subtracting the original signal at each instance by the mean value of the entire EMG signal.

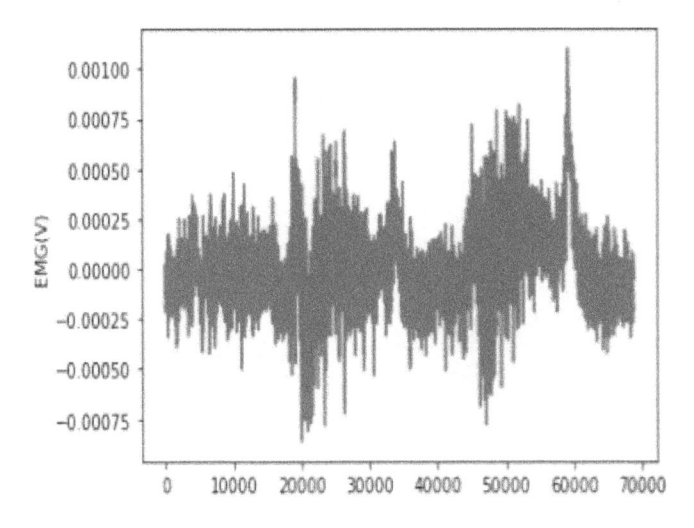

Figure 3.13 Mean corrected signal.

Filtration – The mean-corrected signal is filtered using the Butterworth band-pass filter. The frequency response of the Butterworth filter is flat in the passband and rolls off to zero in the stopband. The rate of roll-off response is determined by the filter order. The nth order Butterworth filter's frequency response is given as,

$$H_{(j\omega)} = \frac{1}{\sqrt{1 + \varepsilon^2 \left(\frac{\omega}{\omega_p}\right)^2}}$$

Where 'n' indicates the order of the filter, $\omega = 2\pi f$, Epsilon ε is maximum pass band gain, (Amax). If we define Amax at cut-off frequency -3dB corner point (fc), then ε will be equal to one. By using the standard voltage transfer function, the frequency response of Butterworth filter is defined as,

$$H_{(jw)} = \left[\frac{v_0(j\omega)}{v_{in}(j\omega)}\right]$$

Where, Vo is voltage of output signal, Vin is input voltage signal, j is square root of -1, and $\omega = 2\pi f$ is the radian frequency. The above equation can be represented in S-domain as given below,

$$H(s) = \frac{v_0}{v_{in}} = \frac{1}{s^2 + s + 1}$$

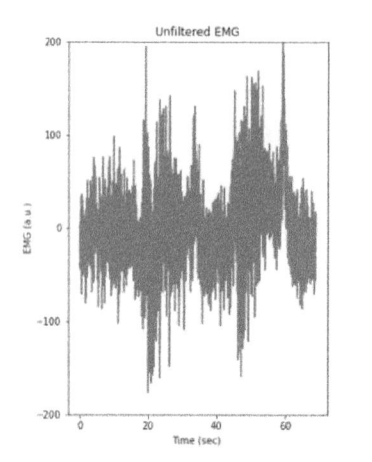

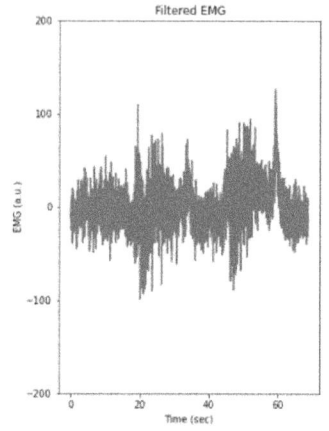

Figure 3.14 Filtered output

The filter used in this case is a Butterworth Filter of bandpass filter of order 4. It is found that the frequency below 30Hz does not hold any significance in pattern recognition, hence the lower cut-off frequency of this bandpass filter is 30Hz. From the FFT, the frequencies of interest lie within 500Hz and above that being high frequency noise. Therefore, the higher cut-off frequency of this filter is 500Hz.

Rectification – It is the conversion of the raw EMG signal to a single-polarity signal. The objective of rectifying a signal is to prevent it from averaging to zero.

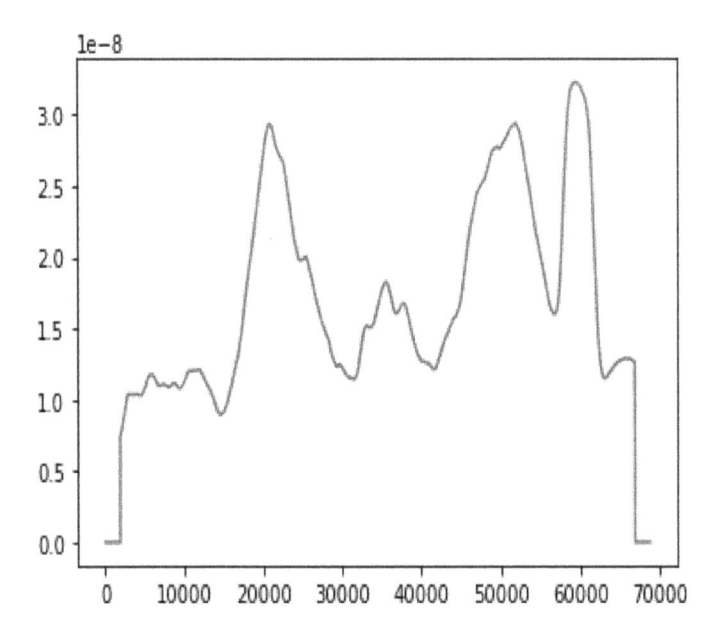

Figure 3.15 Envelope of the rectified signal

3.4.1 Standard Deviation

The standard deviation shows how the EMG signal's value is distributed. It also displays how near each EMG value is to the signal's average value line. The standard deviation value is large if the distance between each data point and the average value is bigger. The standard deviation is determined using the equation below.

$$s = \sqrt{\frac{\sum_{i=1}^{n}(x_i - \bar{x})^2}{n-1}}$$

Here, n is the total length of the EMG signal. $\bar{x}$ is the mean of the signal and x_i indicates the EMG value at any given instant.

Standard deviation = 40.198439

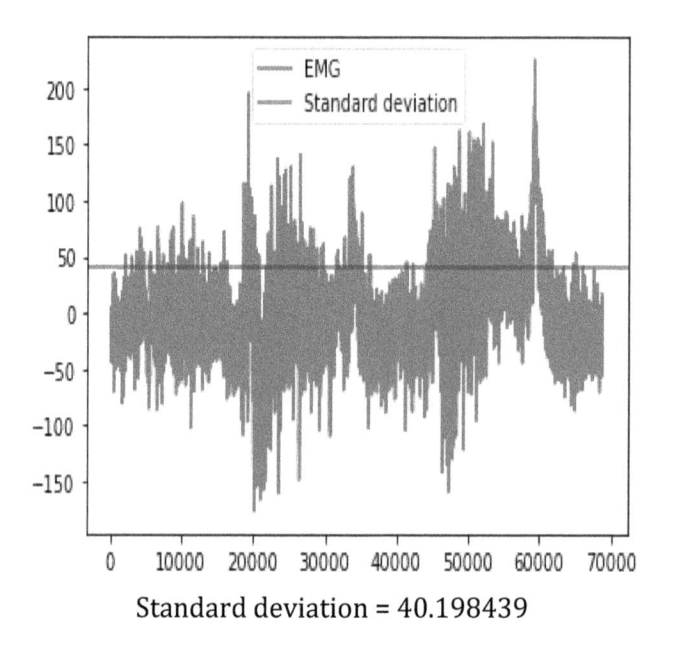

Figure 3.16 *Standard deviation of the entire signal*

Here we see the value close to 50 and subtract that from the noise values.

3.4.2 Frequency Domain Characteristics

With the use of the Fourier Transform, any signal may be mathematically disassembled into a different frequencies collection of sine waves. This usually results in a calculation of each frequency's contribution to the original signal.

The signal being analysed must be steady so as to extract any valuable information from Fourier transform. A stationary signal does not have statistics that change over time. One of the simplest ways to assure signal stationarity is by urging muscle to undertake contact-

Robotics and Modern Computer Vision 43

force and isometric contraction. This is a measurement of how much power each frequency adds to the EMG signal.

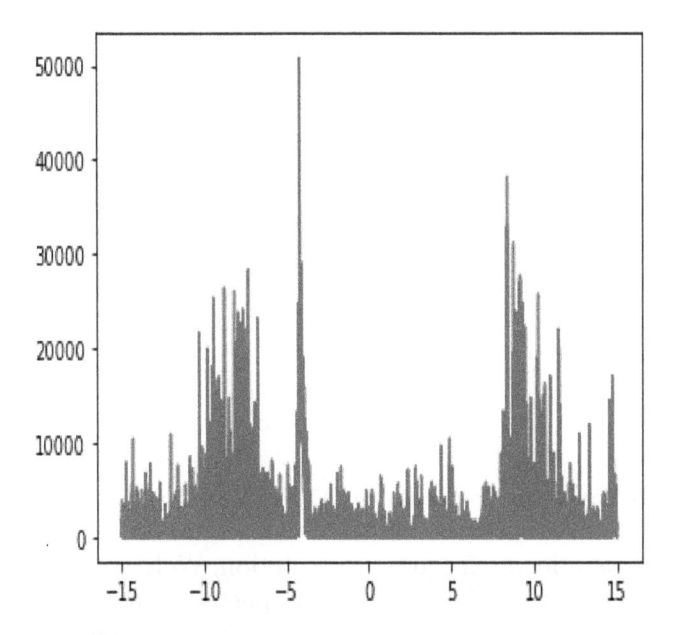

Figure 3.17 Power spectrum.

The frequency components are reviewed throughout the entire length of the EMG data in the power spectral density (PSD). Spectral analysis is based on the examination of frequency component fluctuation across time. The study of muscular exhaustion is the most important use of spectral analysis.

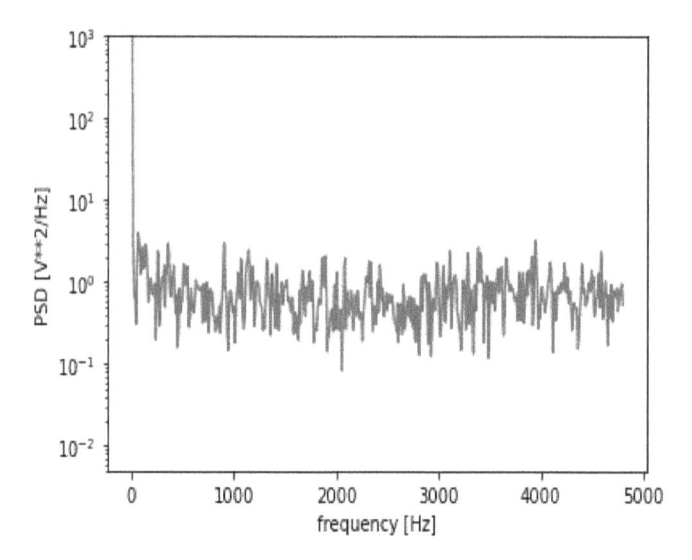

Figure 3.18 Power spectral density using Welch method.

In return for a lower frequency resolution, Welch's technique decreases noise in the calculated power spectra. It improves on Bartlett's method and the usual periodogram spectrum estimation method. Welch's approach is frequently used because it lowers noise caused by finite and defective data. Non-parametric approaches rely on fewer assumptions, such as wide sense stationarily, and so have a considerably broader applicability than parametric methods. The PSD function returns Welch's overlapping segment averaging estimator's PSD estimate of input signal.

When x (the input signal) is a vector, it is considered a single channel. When x is a matrix, each column's power spectral density is computed and stored separately. A one-sided power spectral density estimate is obtained when x is real-valued. A two-sided power spectral density estimate is obtained when x is complex-valued.

The first step in the Welch's method is to divide the signal into the longest possible sequences. The number of sequences obtained should not be exceeding 8, with 50% overlap. Hamming window is used on each sequence. The power spectral density estimate is obtained by averaging the modified periodograms. If the sequence cannot be divided exactly into an integer of 50% overlapping sections, then, we truncate x.

3.4.3 EMG signal's Non-Stationarity

A signal's classification as either stationary or non-stationary can be done based on its timing characteristics. Because of their time-varying features, EMG signals, like the bulk of bio-signals, are classified as non-stationary signals. The problem is to pretend that this nonstationary signal is stationary during small time intervals. This is where usage of stationary analysis techniques is done. However, this particular assumption isn't always valid, necessitating the use of extra non-stationary process techniques.

Many experiments produce multiple-frequency signals that alter in amplitude as they travel through time. The general method is Fourier transform analysis. In contrast, Fourier transform analysis only gives genuine spectra for stationary signals. For signals that change their properties over time, a method of analysis that offers both time and frequency changes is necessary, and one of the most viable approaches is the wavelet transform.

The purpose of this paper is to demonstrate the advantages of wavelet transform over Fourier transform when investigating signals that change their properties over time.

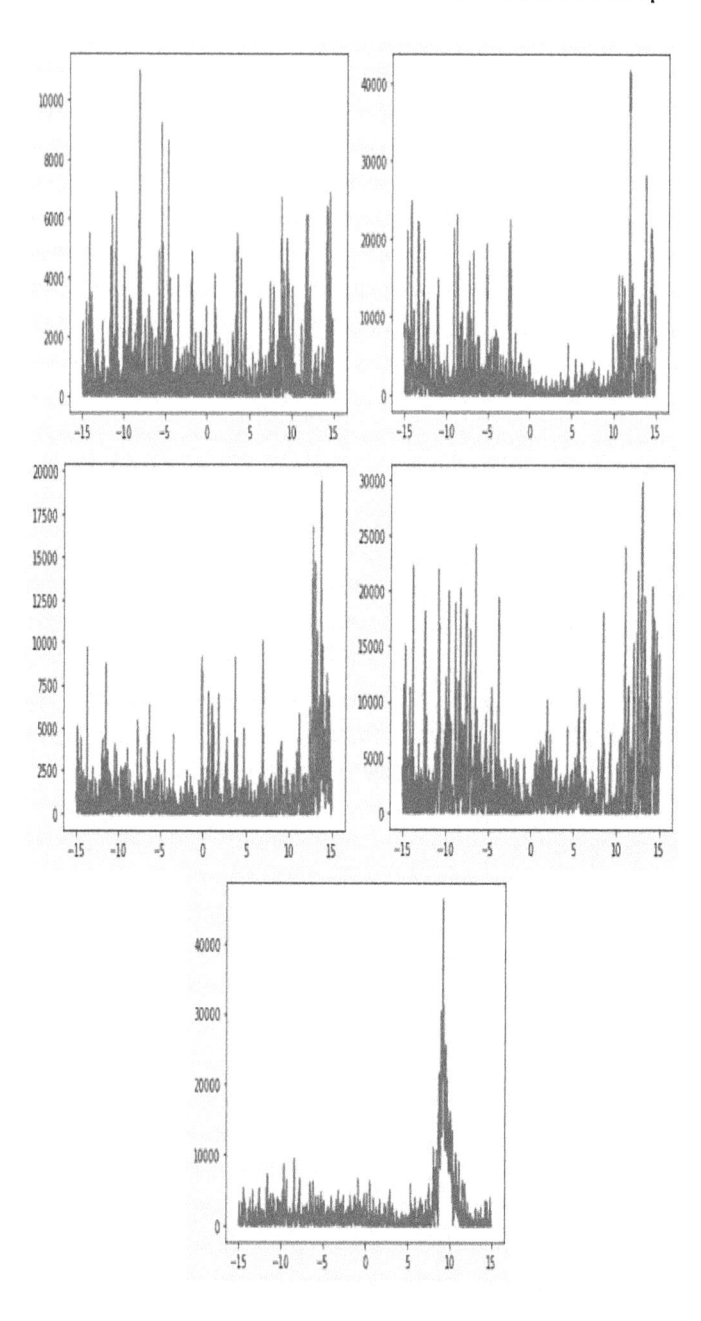

Figure 3.19 *FFT of five segments of the sEMG signal.*

3.4.4 Time-Frequency Analysis

Discrete Wavelet Transform

The DWT transforms the EMG signal with a suitable wavelet basis function, just as a traditional time-frequency analysis (WF). Muscle tiredness is typically assessed using frequency domain analysis of changes in electromyographic (EMG) signals over time. Although Fourier analysis implies signal stationarity, which is improbable during dynamic contractions, it is commonly utilised in these assessments.

Wavelet-based signal analysis approaches do not require stationarity and may be more suited to combined time-frequency domain analysis. The WT's key advantage is that it generates a useful subset of the signal's frequency components or scales.

A wavelet is a waveform with an average value of zero and an effectively limited duration. The signal is viewed on a time scale using wavelet analysis. Wavelet coefficients are the output of a wavelet transform. By multiplication of every coefficient with wavelet which is correctly scaled and shifted, constituent wavelet of the original signal is obtained.

$$[W_\psi f](a,b) = \frac{1}{\sqrt{|a|}} \int_{-\infty}^{\infty} \overline{\psi\left(\frac{x-b}{a}\right)} f(x)dx$$

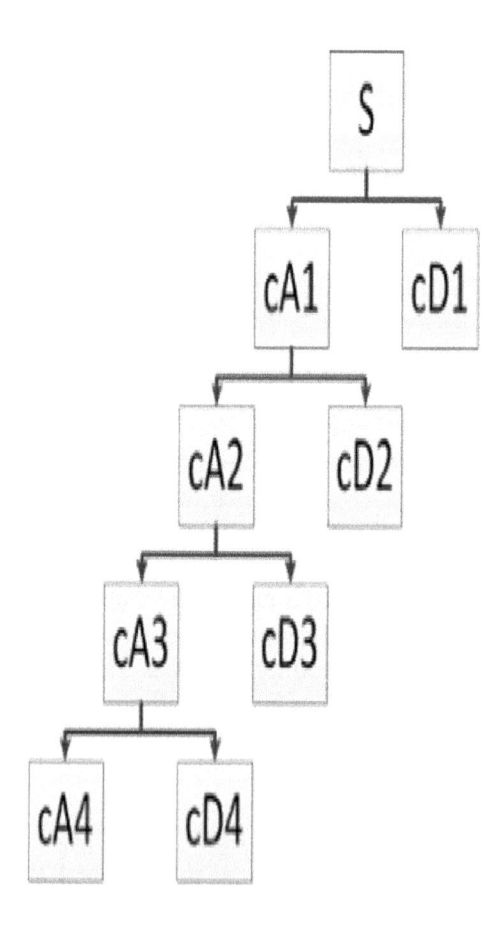

Figure 3.18 *Discrete Wavelet Transform decomposition tree.*

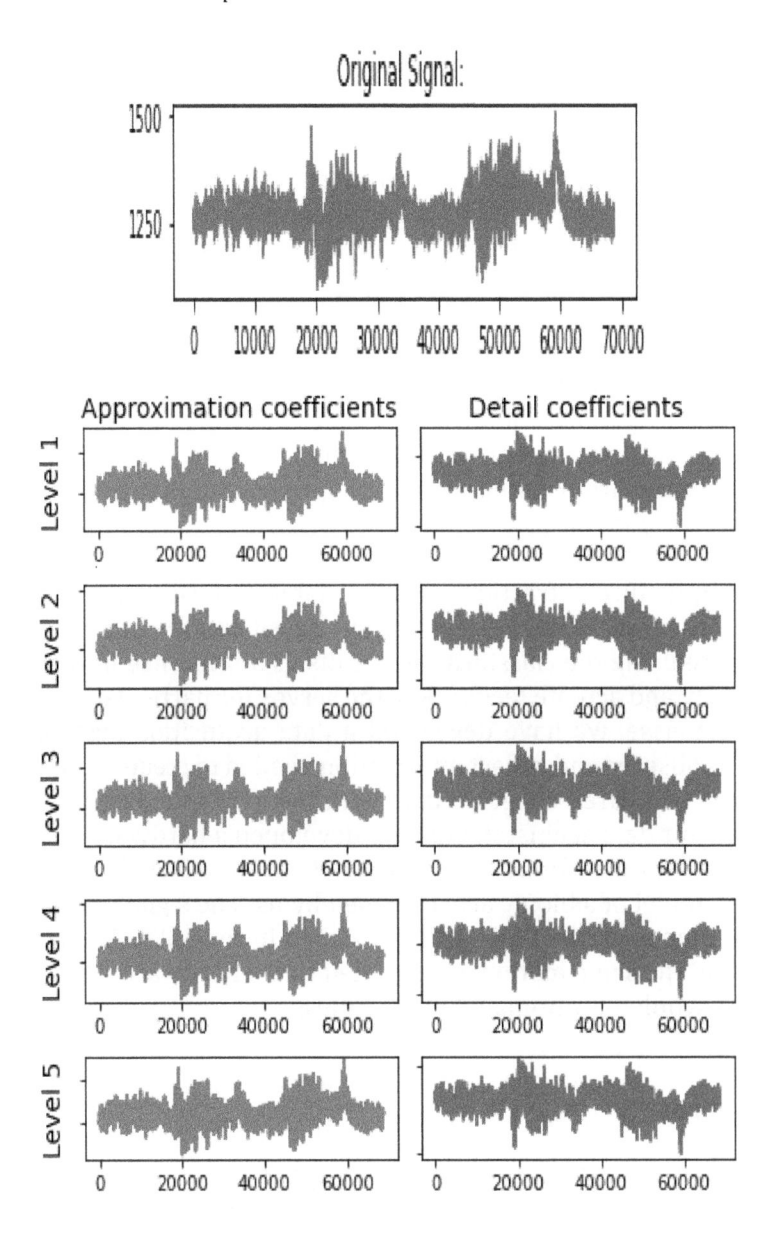

Figure 3.20 Discrete wavelet transform decomposition.

The feature vector's dimensions must be reduced without sacrificing classification accuracy. Furthermore, the dimensionality reduction strategy can improve both the speed and accuracy of the classifier.

3.5 Conclusion

We observed that there were spikes in the signal every time a muscle activity was detected. This helped us verify the actions from available datasets. The time domain and frequency domain signal analysis helped us to narrow down on the prominent features that can be extracted from the signal which can further help us in classification.

The real time mapping of the servo actuation acts as a verification that when a reference signal from the final classification model is used with respect to real time sensor values, the servo can be moved to the particular angle accordingly in the 3D printed prosthetic arm. After repeated trial and error of the signal acquired, we can make an entire dataset from different muscle groups simultaneously using multiple sensor boards.

Applying feature extraction on these datasets based on the previously observed results followed by classification. Implementing hardware using the classification results as a reference in real time to identify and actuate feedback of the servos for particular gestures. To summarise, we have developed a data acquisition system that helps us observe and collect information needed to create a database. We have compared collected data with the available databases and verified our data against it. We have developed a hardware model of the prosthetic arm for a single function that would further be housed in a 3D model of a bionic arm designed by us. The signal acquisition in real time is verified by a GUI developed in the MATLAB environment and the same motion is translated to the hardware. This entire process would be automated in the future.

References

1. Atzori, M., Gijsberts, A., Kuzborskij, I., Elsig, S., Anne-Gabrielle Mittaz Hager, Deriaz, O., Castellini, C., Müller, H. and Caputo, B. (2015).Characterization of a Benchmark Database for Myoelectric Movement Classification, in IEEE Transactions on Neural Systems and Rehabilitation Engineering, 23(1), 73-83.
2. Jamal, Z. M. (2012). Signal Acquisition Using Surface EMG and Circuit Design Considerations for Robotic Prosthesis. In (Ed.), Computational Intelligence in Electromyography Analysis - A

Perspective on Current Applications and Future Challenges, Intech Open, https://doi.org/10.5772/52556.
3. Stashuk, D. (2001). EMG signal decomposition: how can it be accomplished and used? Journal of Electromyography and Kinesiology, 11(3), 151–173.
4. Rafiee, J., Rafiee, M. A., Yavari, F., Schoen, M. P. (2011). Feature extraction of forearm EMG signals for prosthetics. Journal of Expert Systems with Applications, 38(4), 4058-4067.
5. Pancholim, S., Agarwal, R. (2016). Development of low-cost EMG data acquisition system for Arm Activities Recognition, 2016 International Conference on Advances in Computing, Communications and Informatics (ICACCI), 2465-2469.
6. Ahmed, R., Halder, R., Uddin, M., Mondal, P. C., Karmaker, A. K. (2018). Prosthetic Arm Control Using Electromyography (EMG) Signal, 2018 International Conference on Advancement in Electrical and Electronic Engineering, 1-4. doi: 10.1109/ICAEEE.2018.8642968.
7. Jali, M. H., Ibrahim, I. M., Sulaima, M. F., Bukhari, Tarmizi, W. M., Izzuddin T. A., and Nasir, M. N. (2015). Feature extraction of EMG signal using time domain analysis for arm rehabilitation Device, AIP Conference Proceedings, 1660(1), 070041-1 – 070041-9.
8. Phinyomark, A., Limsakul, C., Phukpattaranont, P. (2011). Application of Wavelet Analysis in EMG Feature Extraction for Pattern Classification, Measurement Science Review, 11(2), 45-52.
9. Shair, E. F., Ahmad, S. A., Marhaban, M. H., Mohd Tamrin, S. B., and Abdullah, A. R. (2017). "EMG Processing Based Measures of Fatigue Assessment during Manual Lifting. Hindawi BioMed Research International, Volume 2017, Article ID 3937254, 1-12.
10. Atzori, M., and Müller, H. (2015). The Ninapro Database: a Resource for sEMG Naturally Controlled Robotic Hand Prosthetic. Annual International Conference of the IEEE Engineering in Medicine and Biology Society, 7151-7154.

4

Monitoring System for Patients and to Detect Faulty Devices

Ambika Nagaraj, Department of Computer Science and Applications, St. Francis College, Bangalore, Karnataka, India.

Abstract: Internet of Things (IoT) and robotic based system presented in this work is a system which works to build a smart environment. The system has enormous amount of data which is stored over the cloud. The framework consists of different sensors consisting devices working together to accomplish the task. The readings generated by the sensors are used by robots in its activity. The sensing devices are installed to collect vital signs of the patient and store it over the cloud. The robot is embedded with several algorithms to perform the task of giving medicine to the patient, etc. The robot can be monitored and the data obtained can be used for analyzing robot activities. It can be used to cross-verify their doings and measure the performance. If this system is faulty, then it could affect the patient's health. Hence there should be security measures applied. The system provides security to data by 4.84% and the faulty devices are detected at an early stage by 8.94% compared to the system without any security measures.

4.1 Introduction

Robotics [4][5] is a large field of investigation. Its progress proceeds to increase exponentially every year. Humanoids and mechanization can be a combination of hardware and software. Some of the applications areas are

4.1.1 Healthcare - The robotics can be used for assisting doctors in certain work [8].

4.1.2 Education - There can be a huge impact of robot in education [27][24]. There are consideration levies on to how robot based teaching can help in education area and how its usage instructional humanoids can be better blended into the experiences of students. With the constant appearance of technology, it is worthwhile to recognize the potential of robot as powerful add-ons to knowledge.

4.1.3 Industry – The devices are used in industry sector in certain processing activities. The authors [16] consider companies with smart devices that are part of a shared composition course that requires a portable robot in order to offer the parts they present either to collect them or to take them to another device for more processing.

4.1.4 Agriculture – The devices [29] are used to producegood yield. This module [6] combines a standardized actuator into BoniRob which is an agricultural robot. A penetrometer covers the earth's characteristic steps underneath to pitches of around 80 cm. The number and the conditions of the calibration scores are determined before an automated run. The structure uses devices within the module and GPS. The standard relationships of a penetrometer are the electronic arrangement. It shows consistent properties and provides replications for a considerable quantity of measures. A software ecosystem [15] has been produced to remotelyguide the lawnmower shipping, having a visible indication of the displayed trajectory according to the purposes of the task to be accomplished: scattering, picture recovery, irrigating, breeding, or earth sampling. The route prepared in an isolated node as a series of neutral positions to be reached is sent to the carrier andperformed onboard. The presented track can be visualized at the isolated connection and the production estimated by a collection of heuristics that regulate the achievement in transferring the set of standard states that comprise the complete trajectory. Interaction within the android and the isolated user connections isachieved via Radio-Ethernet by the flying IAI intra- net. The design product [13] is an uncertain planar resemblance organization. It permits us to separate away crop- specific features such as color, spacing, and periodicity. The new process selects the course of the imperative parallel strategy from a simulated illustration. It uses to track the parallel offset of the carrier. It determines the orientation of the camerato guarantee above look is accurate and permanent.

4.1.5 Space - Robotic controllers [20] are mounted on afree-flying satellite. They are envisioned for assembling, sustaining, replacement, and predicament services in the location. These involve proposing in the appearance of non-holonomy, purposing during the planetoid advance stage, and demand preparation during the acquisition of a spacecraft by a multi-manipulator arrangement. The software design [31] uses PolyBot G2. It is based on a supervisor/worker structure. Theadministrator measures the carriage stand and downloads this report to all the section modules on the fly. The division modules act as drudges that solelyadminister the carriage schedule once accepted. PolyBot G2 is examined over restriction programs that require such steps as shifting, modifying, modifying the velocity and amplitude of the sinusoid pace, and shifting from a route to a reptile.

4.1.6 Elderly care - It is an installed method that uses a PIC microcontroller. It implements smart power conservation. It can manage and automate most of the address devices through a flexible smartphone-based android interface. The elements are compared to the enclosed micro-web host within LAN or WiFi modulefor locating, controlling, and regulating things and machines using android-based smartphone utilization. The method also maintains a record of the status of the things.

The work uses robots in sensor system. A sensor system is comprised of a large number of sensing devices. They are densely used either inside the happening orvery adjacent to it. The location of sensor connections need not be superintended or decided. It provides arbitrary deployment in catastrophe assistance transactions. It means that sensor interface rules and algorithms must maintain self-organizing skills. Another unique feature of sensor channels is the collaborative venture of sensor links.Sensor joints come with an onboard processor. Instead of transferring the raw knowledge to the connections qualified for the blending, they use their processing capabilities to narrowly send out uncomplicated estimates and forward only the neededand partly prepared information. Some of its characteristics are:

- Fault tolerance - Some sensor connections may break or be hindered due to a shortage of power, substantial corrosion, or environmental intervention. The breakdown of sensor links should not alter the overall responsibility of the sensor system. It is the authenticity or error threshold concern. Error license is the strengthto maintain sensor system functionalities without any suspension due to sensor linkmalfunctions. The authenticity or liability understands of a sensor connection. It ismodeled by using the Poisson pattern to achieve the possibility of not having a breakdown within the period.

The dilemma [12] of victim disclosure varies from before studied intricacies in shared sign discovery because of the appearance of defects that require specialized processing of the knowledge. Theobstacle also deviates from earlier studied obstacles in understanding such that connections sharing learning may contain restricted eruditon that can be distinct from one link to another. In destination exposure, joints nearby are the objective statement are high-pressure measure. The connections remote from the objective description moderate power densities.

- Scalability - The number of sensor connections in a sensor arrangement can be several degrees of size more than the joints in an ad hoc arrangement. The joints are densely arrayed. The number of devices stationed in investigating an event maybe on the scale of numbers or thousands. Depending on the administration, the estimate may lead to an excessive amount of millions. New designs must be able to operate with this abundance of connections. They must also employ the high frequency of the sensor interfaces.
- Cost - The arrangements contains of a huge number of sensor joints. The payment of an individual connection justifies the overall value of the arrangement. The priceof the arrangement is expensive. The sensor system is not cost-justified. As a result, the cost of each sensor node has to be kept low. The state-of-the-art technology provides a Bluetooth communication method to be of low cost. The valueof a pico node is targeted to be less than one dollar. The payment of a sensor connection should be much less than one dollar. The sensor system is achievable. The value of a Bluetooth receiver is known to be a low-cost design.
- Hardware constraints - A sensor connection has four essential elements. They mayalso have application-dependent ingredients such as a location finding method, energy generator, and mobilizer. Sensing assemblages comprise twosubunits- sensing devices and analog-to-digital converters. The analog signs delivered by the sensors based on the discovered happening are switched to digitalby the ADC and then stuffed into the processing assemblage. The processing unit connects with a miniature accommodation. It regulates the methods making the sensor connection cooperate with the other joints. It carries out the designated sensing duties. A transceiver assembly attaches the junction to the system. One ofthe most ingredients of a sensor

connection is the power complement. Power associations establish by strength scavenging such as cosmic blocks. Other subunits are application-dependent. Most of the sensor arrangement routing procedures andsensing duties need an understanding of position with high exactness. The sensor joint has a discovering decision arrangement. A mobilizer needs to move sensor connections to carry out the distributed responsibilities.

- Power consumption - The broadcast sensor connection is a microelectronic design.

It is equipped with an insufficient strength reservoir. In some application scenarios,replenishment of strength sources might be impracticable. Sensor attachmentlifetime shows a specific connection to battering lifetime. In a multihop ad hoc sensor system, each connection plays the twofold function of data originator and information router. The malfunctioning of several joints can cause notable topological transitions and might require rerouting of packages and reorganizationof the system. Hence, strength maintenance and administration take on additionalsignificance.

The mechanical design [7] expands anddevelops the abilities of its principles during the presentation of novel technologiessuch as digitization that converts the application to an interconnected global arrangement in which devices, methods, and products communicate together. It deals with the portable humanoid travel problem in technical circumstances wherethe arrangement rarely encounters major alterations.

In this work, the sensing devices are used as wearable's in the patient environment. The robots are used to provide medicines to the sick and share the same information with the sensors. These sensors are used as authenticatingdevices. They authorize the automaton [17] [3], collect the respective data and share the processed data with the host machine. The chapter is divided into five sections. Literature survey is detailed in section two. Third segment explains the proposed work. The fourth paragraph describes the simulation and its analysis. The work is concluded in fifth section.

4.2 Literature survey

Some of the contributions given to the domain of healthcare is discussed in this section. The work [26] is a study of Parkinson disease-affected personalities. The data is collected from information with the aid of portable receivers in the healthcare trade and identification of the motion representation of Parkinson disease-affected personalities. The robot preprocesses with adequate information set to understand the manifestation of all conditions. Robot activities define the path. It detects restriction areas, protected areas, and dangerous areas. It operates based on guided knowledge. If there is an inaccuracy in the information, it may cause severe destruction during arrangement processing. The laser chain finder examines the way for the pedestrian change of both the robot and the client. The laser chain scanner is improved for road situation analyzers to recognize the obstruction. The knowledge- defining procedure uses the laser range scanner method initially guides the Artificial Intelligence with dimension andinvestigates the carriage achievement. The laser scanner provides a field of design screening the complete 180° flounces using the laser radar. The method of the laser radar commences from theright and terminates at the left. The aloofness and position of the material with the robot's path depending on the beam divergence and resolution set. It tackles the restriction of nearby robots. The laser discovery layer has subgroups. It falls in the protected precinct and is in the middle of the automaton. The sensing device is in the robot handle, and another mini detector is determinedin the user's leg to investigate the motion representation. It is irregular in its movements. FOG doesnot know operative stepping. The FOG is practice on the platform. Since this disorder is permanent, researchers show enthusiasm toward the disease in the health care industry. These effects use knowledge examination. The knowledge in healthcare management shows a tremendous amount of structured and disorganized information for motion representation. Evaluation of FOG is the most important for recognizing the signs. The motion analyzer observes the motion variability according to the expedition, speed, and step interval. Dynamic time twist proposal decides the time series between client and automaton. The path is planned using model form.

The primary monitoring [30] of various parameters in the therapeutic department presents the interference of cutting-edge technology

IoT in the domain of healthcare where patients, as well as specialists, can continuously record a patient's well-being. The series of pre-planning and preparation of complicated surgeries gives recommendations to functional consultants conducting transactions. Computer-assisted surgery provides surgeons to operate from remote locations. The master tool manipulators and patient sidemanipulators interfaces using IoT. Master tool manipulators constrain patient side manipulator according to the surgeon's data, whereas patient side manipulator is provided with sensory accessories like a power sensor, optics sensor, substantial sensor with various actuators to implement parameters to the physician. With master tool manipulators, doctors feel the haptic feedback, which presents intuitive awareness of the industry conditions inside the sufferer.

The work [9] is structure that allows interconnection and collaboration of various IoT interfaces, multimodal interfaces, and assistance automata. The main controller accumulates knowledge from instruments connected in the smart place. It provides an unrestricted interface to envision and manage them. The Internet of Things (IoT) foundation has various sensor and actuator interfaces interconnected by the same gateway. Assistance automata correlate to the structures using a Wi-Fi interface. Administrator and contributor nodes are in Robot Operating syst3em. The subscriber node is in charge of communication. It has the primary controller. Representational State Transfer requests to get knowledge or to adjust the amounts of the actuators. Other publisher nodes run on the robots. It can publish messages after receiving them from the subscriber node. A multimodal interface allows the user to interact with the environment by different means. It has a tablet PC outfitted with an eye-tracker and a Microsoft Kinect camera. The tablet PC affords substantial authority, sound, and eye-based administration.

The working arrangement [32] of the structure has WBAN, Wireless Neighborhood Area Networks, Wireless Wide Area Networks, and the Datacenter. The material working in each practical segment includes materials like wearable, smart phones, pads, and datawarehouse designs. These tools execute using IPv6 and 6LoWPAN as the system overlay rules. The concrete panel communication uses the IEEE 802.15.4 order. However, since for IoThealthcare, some assistance requires healthcare providers to have complete admittance to the knowledge for the constant monitoring of cases despite their position, real-time and seamless connection is needed. Movement rules for presentation are within the IoThNet and can be afforded by using the 6LoWPAN order for corresponding communications among portable victim machines, base systems, and attended channels. The construction uses the DPWS etiquettepile to describe methods as aids. The plan is to combine and manipulate various heterogeneous fitness and homecare tools in supported maintenance circumstances efficiently and seamlessly. Ambient knowledge arrangements consist of pharmaceutical sensors and accessories,machines, radio channels, and software employment, which are used to recognize stimuli and create knowledge from the users and the surroundings. The information produced enables the monitoring of cases. It facilitates the understanding of Ambient assisted living healthcare applications. Ambient assisted living utilization facilitates user-dependent co-operations for elderly and disabled characters. These user-dependent aids include monitoring the necessary warnings or managing the tracking of patients. They interpret different sensed erudition and detecting and responding to unusual circumstances.

The IoT-based program [28] allows the RFID-based testimony [14] of the devices, questioning and regaining therapeutic data from multiple actual healthcare learning practices. It shows significant knowledge of the approved physiques. The purpose of running on portable machines gives healthcare providers the learning and abilitiesthey need wherever and whenever they need them. This method permits tracing RFID-tagged objects to afford innovative quality assistance for the movement of victims, pharmaceutical personnel, therapeutic equipment, and other things. The strategy assures the positive outpatient description within a pharmaceutical convenience. Furthermore, it extends weak association across medicinal equipment frames, for example, through the use of a specialized agency that achieves specific learning giving rules.

4.3 Proposed system

In healthcare, the approach of staying connected with information about patient's health and monitoring the important vital health parameters of the patient, providing of round the clock support, etc. are some of the necessary things which are always given importance in healthcare. Portable healthcareassistance refers to aids that should not be affected even if patients are on the move and thus need equipment and technologies to assist a patient's movement.

Sensors [1] [21] are used in innumerable applications. Sensors are used for monitor or detecting purposes. For example, they can be used in healthcare related devices like for heart rate monitoring, etc. Sensors are also used for robots. Sensors help the robot to collect data. In this work, the devices consisting of sensor are used for monitoring health of the patient. These are wearable devices. The devices will keep on sensing health parameters of the patient and will continue to monitor patient health. The data collected is sent to host/server. Patients need to be provided with medicines time to time. To go and give medicines to the patients, robots are used. The prescribed medicines by doctors are given by robots to the patients. The robots are first authenticated. Then only they are authorized to go and give medicines to the patient.

Table 4.1 *Notations used in the study. Wearable device consists of sensor. Hence it can be referred to as wearable sensor.*

Notations	Description
N	Network
H	Host/server
S_i	Wearable sensor
R_{id}	Identity of the robot
R_{loc}	Location of the robot
S_{id}	Identity of the sensor
S_{loc}	Location of the sensor
M_i	Medicine details
PK_i	Public key
D_i	Data transmitted by the wearable sensor
ACK_i	Acknowledgement sent by the server

4.3.1 Embedding credentials -Different algorithms and key credentials are used for the robot before initiating it. They help the robot to recognize –
- The patient's w.r.t the distance they are positioned.
- Reading the vitals from the sensors after authenticating themselves.

4.3.2 Moving to the pre-destined position

The robot is given the location information to travel or move to the desired location and provide medicines to the patient. They authenticate themselves using their identification. A hash key is generated and exchanged with the sensors. Equation (2.1) is exchanging the hash key using its identity R_i and location R_{loc} with the sensor S_i. The sensing devices also create a hash key using its identity and location information. Equation (2.2) generates the hash key using its identity S_i and location S_{loc} and transmits it to the robot R_i. After successful authentication, the robot can go and give medicine to the patient.

$$R_i \rightarrow hash(R_{id} || R_{loc}) : S_i \qquad (2.1)$$

$$S_i \rightarrow hash(S_{id} || S_{loc}) : R_i \qquad (2.2)$$

4.3.2 Exchange of data

The robot shares the necessary data w.r.t the patient. The robot shares the data consisting the details about medicines given to the patient. The same is represented in the equation (2.3). The robot R_i shares medicines details M_i with the sensor S_i.

$$R_i \rightarrow M_i : S_i \qquad (2.3)$$

4.3.4 Broadcasting the public key

The Host machine (server) H broadcasts the public key [2] [25] to all the stationary sensors. In the equation (2.4) the host H is broadcasting the public key PK_i with the network N.

$$H \rightarrow PK_i : N \qquad (2.4)$$

4.3.5 Transmitting the processed data with the server

A public encryption algorithm to encode the processed contents broadcasted by the server is used. The hash key is also attached to the transmitted message. In the equation (2.5) the data D_i is encrypted using the publickey PK_i. The hash key is attached to the encrypted data and transmitted to the host H.

$$S_i \rightarrow PK_i(D_i) || hash(S_{id} || S_{loc}) : H \qquad (2.5)$$

4.3.6 Acknowledging the message

The last sensed data is kept till there is theacknowledgement received from the host machine. After receiving the acknowledgement message, the memory of the last sensed data is refreshed. In equation (2.6) the host is sending the acknowledgement ACK_i to the sensor S_i.

$$H \rightarrow ACK_i: S_i \qquad (2.6)$$

Table 4.2 *Algorithm used to generate the hash key*

Input parameters used – identification of the device (16 bits, location information (32 bits)
Step 1: Input the identification and location details
Step 2: divide the 48 bits into 4 parts.
Step 3: Apply AND operations on 1 part and 3 parts. The outcome results in 12 bitsStep 4: Apply OR operation on 2 part and 3 part. The outcome results in 12 bits
Step 5: Apply XOR operation on 1 outcome and 2 outcome (resultant – 12 bits)

4.4 Analysis and simulation

Small wearable devices are used for the patients. These devices sense the vitals of the patient and store them in their memory. They process these data, encrypt them and transmit them to the server. Robot is used in the work to provide medicines to the patients. It can even go and give the prescribed medicines by doctors to the patients. The robots are fed withalgorithms that assist them to move to the allotted location and deliver the medicine to the patients.

The work is simulated using NS2. The robot is considered as mobile node in the simulation. 6 static nodes installed reflect the patient location. Table 4.3 contains the description of all the parameters used in the simulation.

Table 4.3 Parameters used in the simulation

Parameter	Explanation
Network dimension	200m * 200 m
Number of Static nodes installed (depicting patient location)	6
Number of mobile nodes used (depicting robot)	1
Node installation	Manual
Length of the hash key used	12 bits
Length of data transmitted by sensor to the server	1024 bits
Length of data transmitted by robot	254 bits
Simulation time interval	60s

4.4.1 Authentication

Authentication [10] is essential before commencingthe actual communication. The contribution uses mutual authentication procedure [22]. The robot needs to authenticate first. After that the details of the medicine are given to the patient. It authenticates itself by generating hash key using its identification and location details. The sensing devices also authenticate by generating hash key [18] using its identification and location details. This procedure aims to provide security to the system. Securing [11] the communication using encryption keys provides additional security to the data. The framework provides 4.84% more security to data compared to system without any security measures. The same is represented in Figure 4.1.

Robotics and Modern Computer Vision

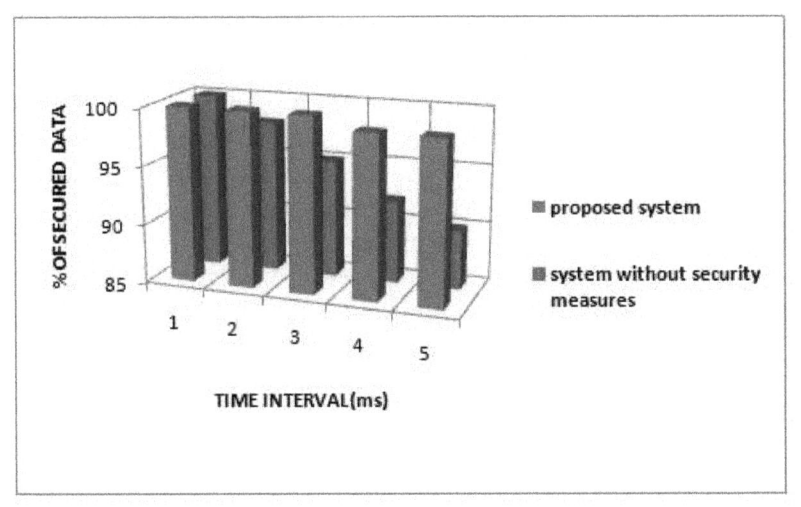

Figure 4.1 Security to data

4.4.2 Detecting the faulty devices

The devices if faulty [19] can be detected at an early stage, if the devices are authenticated. The faulty devices are detected at an early stage by 8.94% better than the system without security measures. The same is depicted in Figure 4.2.

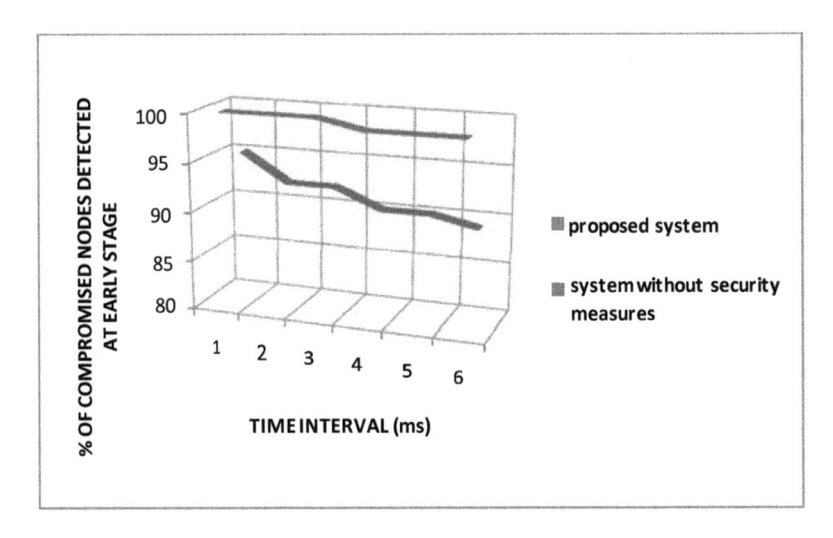

Figure 4.2 Detection of faulty devices at early stage.

4.5 Conclusion

The healthcare infrastructure can benefit from automation. The system presented in this chapter can be used to monitor the patient's health. It can even go and give the medicines which are prescribed by doctors to the patients. The system uses robot to deliver the medicine to the patients. The system requires some security measures that track the system. The proposed work monitors and tracks the activities of the robot before delivering the medicines to the patient. The wearable devices and the robot are authenticated. The server broadcast the public key to the wearable devices. After successful authentication, the processed data is transmitted to the host machine/server. The system provides security to data by 4.84% and the faulty devices are detected at an early stage by 8.94% compared to thesystem without any security measures.

References

1. Ambika, N. (2020). SYSLOC: Hybrid Key Generation in Sensor Network. In Singh. P., Bhargava. B., Paprzycki, M., Kaushal, N., Hong, WC. (eds), *Handbook of Wireless Sensor Networks: Issues and Challenges in CurrentScenario'., Advances in Intelligent Systems and Computing*, vol 1132. Springer, Cham, pp. 325-347. https://doi.org/10.1007/978-3-030-40305-8_16.
2. Ambika, N., & Raju, G. T. (2010). Figment Authentication Scheme in Wireless Sensor Network.*Security Technology, Disaster Recovery and Business Continuity* (pp. 220-223). Jeju Island, Korea: Springer, Berlin, Heidelberg.
3. Argall, B. D., Chernova, S., Veloso, M., & Browning, B. (2009). A survey of robot learning fromdemonstration. *Robotics and autonomous systems, 57* (5), 469-483.
4. Armesto, L., Fuentes-Durá, P., & Perry, D. (2016). Low-cost printable robots in education. *Journal of Intelligent & Robotic Systems, 81* (1), 5-24.
5. Balch, T., Summet, J., Blank, D., Kumar, D., Guzdial, M., O'hara, K., et al. (2008). Designing personal robots for education: Hardware, software, and curriculum. *IEEE Pervasive Computing, 7* (2), 5-9.
6. Bangert, W., Kielhorn, A., Rahe, F., Albert, A., Biber, P.,

Grzonka, S., et al. (2013). Field-robot-based agriculture: "RemoteFarming. 1" and "BoniRob-Apps. *VDI-Berichte*, 439-446.
7. Benotsmane, R., Dudás, L., & Kovács, G. (2018). Collaborating robots in Industry 4.0 conception.*IOP Conference Series: Materials Science and Engineering* (pp. 1-10). Kecskemét, Hungary: IOPPublishing.
8. Bogue, R. (2011). Robots in healthcare. *Industrial Robot: An International Journal.*
9. Brunete, A., Gambao, E., Hernando, M., & Cedazo, R. (2021). Smart Assistive Architecture for the Integration of IoT Devices, Robotic Systems, and Multimodal Interfaces in HealthcareEnvironments. *Sensors, 21* (6), 1-25.
10. Burrows, M., Abadi, M., & Needham, R. M. (1989). A logic of authentication. *Proceedings of theRoyal Society of London. A. Mathematical and Physical Sciences* (pp. 233-271). Royal Society ofLondon.
11. Chan, H., & Perrig, A. (2003). Security and privacy in sensor networks. *Computer, 36* (10), 103- 105.
12. Clouqueur, T., Saluja, K. K., & Ramanathan, P. (2004). Fault tolerance in collaborative sensor networks for target detection. *IEEE transactions on computers, 53* (3), 320-333.
13. English, A., Ross, P., Ball, D., & Corke, P. (2014). Vision based guidance for robot navigation in agriculture. *IEEE International Conference on Robotics and Automation (ICRA)* (pp. 1693- 1698).
14. Finkenzeller, K. (2010). *RFID handbook: fundamentals and applications in contactless smart cards, radio frequency identification and near-field communication.* Hoboken, New Jersey: John wiley & sons.
15. Garcia-Alegre, M., Ribeiro, A., García-Pérez, L., Martínez, R., Guinea, D., & Pozo-Ruz, A. (2001). Autonomous robot in agriculture tasks. *3ECPA-3 European Conf. On Precision Agriculture*, (pp. 25-30). France.
16. Gonzalez, A. G., Alves, M. V., Viana, G. S., Carvalho, L. K., & Basilio, J. C. (2017). Supervisorycontrol-based navigation architecture: a new framework for autonomous robots in industry 4.0 environments. *IEEE Transactions on Industrial Informatics, 14* (4), 1732-1743.
17. Goodrich, M. A., & Schultz., A. C. (2008). *Human-robot interaction: a survey.* Norwell, MA: Now Publishers Inc.
18. Guesmi, R., Farah, M. A., Kachouri, A., & Samet, M. (2016).

Hash key-based image encryption using crossover operator and chaos. *Multimedia tools and applications, 75* (8), 4753-4769.
19. Mathews, M., Song, M., Shetty, S., & McKenzie, R. (2007). Detecting compromised nodes in wireless sensor networks. *Eighth ACIS International Conference on Software Engineering, Artificial Intelligence, Networking, and Parallel/Distributed Computing* (pp. 273-278). Qingdao,China: IEEE.
20. Moosavian, S. A., & Papadopoulos, E. (2007). Free-flying robots in space: an overview of dynamics modeling, planning and control. *Robotica , 25* (5), 537-547.
21. Nagaraj, A. (2021). *Introduction to Sensors in IoT and Cloud Computing Applications.* UAE: Bentham Science Publishers.
22. Otway, D., & Rees, O. (1987). Efficient and timely mutual authentication. *ACM SIGOPS Operating Systems Review. 21*, pp. 8-10. New York,NY,United States: ACM.
23. Qadri, Y. A., Nauman, A., Zikria, Y. B., Vasilakos, A. V., & Kim, S. W. (2020). The future of healthcare internet of things: a survey of emerging technologies. *IEEE Communications Surveys & Tutorials, 22* (2), 1121-1167.
24. Saerbeck, M., Schut, T., Bartneck, C., & Janse, M. D. (2010). Expressive robots in education: varying the degree of social supportive behavior of a robotic tutor. *Proceedings of the SIGCHI conference on human factors in computing systems* (pp. 1613-1622). New York,NY,United States: ACM.
25. Salomaa, A. (2013). *Public-key cryptography.* Berlin/Heidelberg, Germany: Springer Science & Business Media.
26. Sivaparthipan, C. B., Muthu, B. A., Manogaran, G., Balajee Maram, R. S., Krishnamoorthy, S., Hsu, C.-H., et al. (2020). Innovative and efficient method of robotics for helping the Parkinson's disease patient using IoT in big data analytics. *Transactions on Emerging Telecommunications Technologies, 31* (12), 1-11.
27. Toh, L. P., Causo, A., Tzuo, P. W., Chen, I. M., & Yeo, S. H. (2016). A review on the use of robots in education and young children. *. Journal of Educational Technology & Society, 19* (2), 148-163.
28. Turcu, C. E., & Turcu, C. O. (2013). Internet of things as key enabler for sustainable healthcare delivery. *Procedia-Social and Behavioral Sciences. 73*, pp. 251-256. ELSEVIER.
29. Vasconez, J. P., Kantor, G. A., & Cheein, F. A. (2019). Human–

robot interaction in agriculture: A survey and current challenges. *Biosystems engineering, 179*, 35-48.
30. Verma, V., Chowdary, V., Gupta, M. K., & Mondal, A. K. (2018). IoT and robotics in healthcare.In A. E. Hassanien, N. Dey, & S. Borra (Eds.), *Medical Big Data and Internet of Medical Things* (pp. 245-269). London: CRC Press.
31. Yim, M., Roufas, K., Duff, D., Zhang, Y., Eldershaw, C., & Homans, S. (2003). Modular reconfigurable robots in space applications. *Autonomous Robots, 14* (2), 225-237.
32. Zeadally, S., & Bello, O. (2019). Harnessing the power of Internet of Things based connectivity to improve healthcare. *Internet of Things*, 1-14.

5

Early Stress Identification and Detection from Facial Expressions

Sheema Sadia, Apurva Kumari and M. C. Chinnaiah
Department of Electronics and Communication Engineering, B V Raju Institute of Technology, Narsapur, Hyderabad, Telangana, India

Abstract: Stress is a major problem in our society, as it is the source of many health problems. This work concentrates on this problem which confronts everyone today. Stress will have adverse effects on people if it is not detected at an early stage. In this work, human emotions with monitored facial expression sense stress as a function of the captured image. An approach based on convolutional neural network (CNN) was introduced. Here, the considerable emotions are happiness, sadness, frustration, surprise, disgust, apprehensive, neutral. Thus, deep learning has an ability to understand the characteristics that will allow machines to generate perception. The emotional analysis performed at the end of each iteration suggests that reducing the invasive nature of the device can influence user perceptions and improve classification performance.

5.1 Introduction

Mental stress can cause due to some illness related to emotions which can further lead to cardiovascular diseases and other risk related to life threat so it should be detected at very early stages. Many methods are there to detect stress such as biomedical, self-reporting questions but detecting with the help of this method will take long time which is not so easy. Basically, emotions play a vital role in stress as emotions are related to stress so emotions canhelp us to understand stress one individual going through. The work in this chapter will be helpful for detecting stress at early stages. Nowadays, deep learning is becoming very popular in very various streams of industries and also make life easier. Machine Learning is a method of design of programs that can learn from codes, easily adapts changes, and upgrade execution with experience. Computers are expected to solve complex complications and is widely used in our daily lives.

Humans use senses to adopt insights of the environment. So, machine perception aims to imitate senses of human in order to interconnect with the environment. Machines have different ways to capture the environment by using cameras or sensors. Therefore, by using the information with relatedalgorithms which permit to produce machine perception. Deep learning (DL) algorithms has been shown to be truly demonstrated for successful. [1] proposed convolutional neural network for classification of videos.

Automatic speech recognition, natural language processing, and computer vision are few of the disciplines where DL had success. One of the key advantages of employing DL methods is that there is no need of engineer. [2] presented a method of detecting the facial traits of stress and/or anxiety. [3] put forward an optimization algorithm for learning. [4] presented an expression of facial model for facial detection. Above fundamental representations, algorithms can learn traits about their own. For example, pixel representations of images can be fed into an ANN for image recognition. The program will then determine whether a certain combination of pixels constitutes a recurrent feature of the image. The characteristic expression utilizing LBP and deep learning was presented in [5]. The authors in [6] introduced the detection of negative facial emotions using a connected convolution network.

When employing Stochastic Gradient Descent, using ReLU in Figure 5.1 reduced the number of epochs required to converge by a factor of 6. ReLU's fragility if the inputs distribution is below zero then it is a significant disadvantage. Because of practical considerations, GPU training has become essential for training deep networks.

$$f(u) = \max(0, u)$$

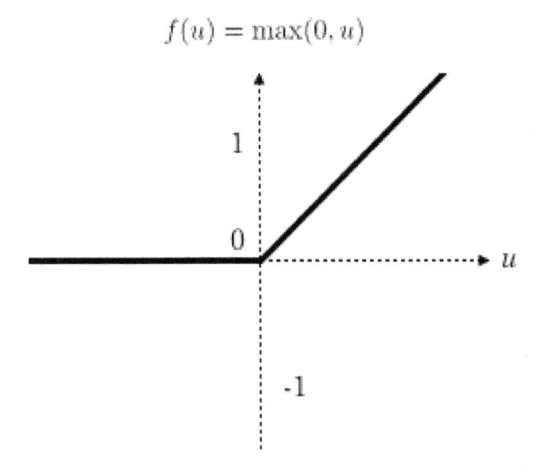

Figure 5.1 *ReLU model.*

5.2 Facial Emotions Identification and Recognition

Facial emotions transfer data related to inner state. Machine has ability to generate sequence of images. The further use of deep learning techniques will be helpful for machines to get idea of their internal mood. Deep learning has an ability to become a important factor to order to create interaction between both humans and machines. When supplying machines with different kind of self-awareness with respect to human peers on how to work onimprovement its interaction with natural intelligence. Deep learning make use of hierarchical neural network to recognize related data. Neuron codes will be bounded together inside this hierarchical neural networks which is close to the human brain. Other traditional methods in machines are hierarchical structure nonlinear approach and process of data within a series of layers then it will integrate subsequent layers additional data. The structure of the neural connected network is related to the structure of the human brain. In Figure 5.2, the different emotions identification can be done using feature extraction and classification. To achieve theaccuracy of detection of emotions convolutional neural network has been used.

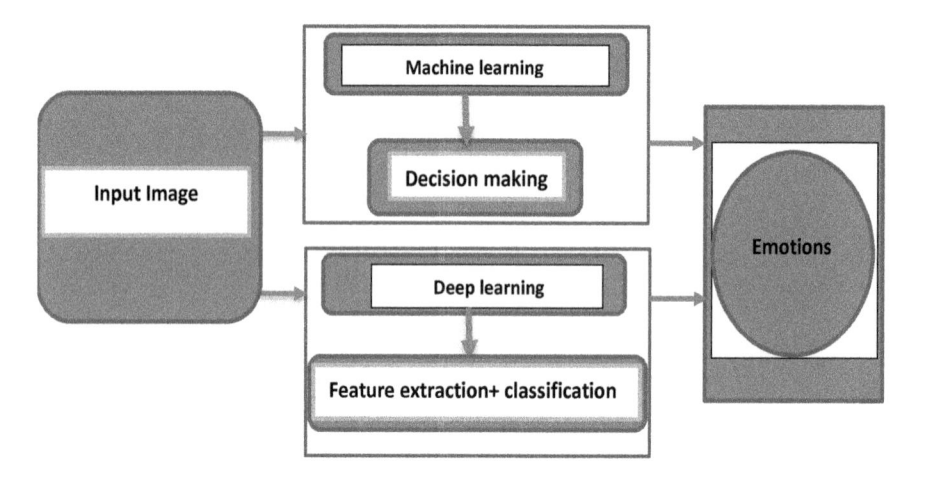

Figure 5.2 Different emotions analysis through feature extraction and classification

We utilize brain to recognize patterns and differentiate various types of information, neural networks can be used to do the same activities on data. Each individual type of layers of neural networks can be a filter which works from initial gross to end, increasing from possibility of detecting then giving an output of result. Our human's brain similar to that work. Whenever it receives new type of information, then brain starts to compare them with the familiar objects. Then the same concept is used by deep neural networks. Neural networks are capable to do many activities, such as regression, classification and clustering. By the help of neural networks, one can sort unlabelled data based upon the similarities. Or else in the caseof recognition, one is capable to train upon a network within a labelled to recognize the various samples in the data.

A CNN is a deep learning method that will extract an input image and start giving importance (biases and weightable) to different aspects of objects in the image, with anability to distinguish between them. Figure 5.3 shows the convolutional neural network architecture for detecting the different emotions. When it is being compared with other classification technologies, the few amounts of pre-processing essential for ConvNet is usuallyless. But basic approaches require handmade engineering filter. ConvNets can used to learn the characteristics of filters with essential training. The structure of a ConvNet is bought by cortex and is skin to the pattern of connectivity in neurons of the human Brain.

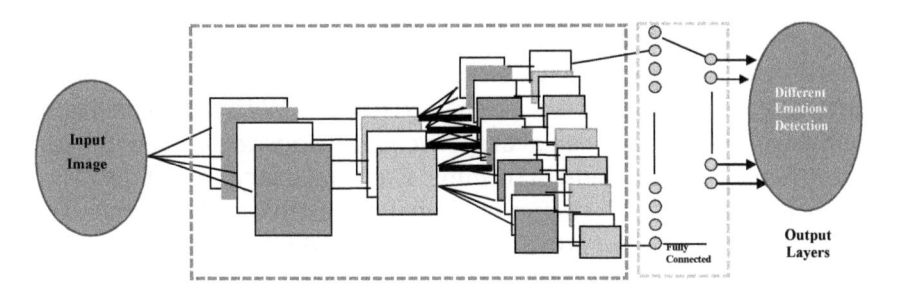

Figure 5.3 *Convolutional neural network*

With the use of sustainable filters, a ConvNet will successful capture the temporal and spatial dependencies of the image. Due to the number of reduced parameters used in the process. The reusability of weights and the structure performs super fitting of the image. In other words, the type of network should be trained identify image s sophistication. An RGB image has been divided into its three colour planes like red, green, and blue. Images can be restored in a type of colour, including Grayscale, RGB, CMYK.The ConvNet'sis to compress the format of image athat is easy to process when elements of preserving that is important for a decent obtaining a prediction. This is difficult when designing a structure which is capable of identifying features which is also being scalable to big.

The Convolution Operation goal is to generate high resolution characteristics from theinput image like edges. No need of limit to ConvNets to more than one convolutional layer. The initial layer is literally responsible for capturing low resolution information such as colours, edges, gradient direction. The operation generates different types of results in first case the dimension of the feature is made low when compared with the input and in other case the dimension is either increased or remain constant. This is accomplished by using padding validity in the above case and same padding will be used second case. So one can generate many output images from a single image with several filters in convolution. Then it is used as a horizontal extractor and a vertical extractor to generate two output images for the handwritten digit. We can use a variety of additional filters to generate more feature maps, which are also known as output images.

When you acquire the feature maps from a convolution layer, it is very difficult to add both pooling and sub-sampling layer in this method. The Pooling portion, like CNN Layer, in charge of shrinking convolved feature structural size. The cascade classifier is made up of several stages, each of which is made up of a group of dull learners. The dull learners are conclusion stumps, which are simple classifiers. Boosting is a strategy at each step used to train. With taking decision of the weak learners by an average weighted boosting is used in high classifier accurate to help in training. Each and every step of the classifier is assigned with region of a label of positive and negative described by the present position of sliding window.

The positive value describes that an image was discovered, and a negative value describes that no objects were discovered. In case of the label is of negative value, the detector then shifts window to the further places after completing categorization of the region. When it reaches final phase and qualifies as the positive region, the detector informs an image object present at the present window area. The phases are set up in such a way that negative samples are rejected as quickly as possible. The presumption is that the object of interest is not visible in the very large major part of windows. True positives, on other hand, are uncommon and worth investigating. The experience that the algorithm learns is based on the observations in the training set. Each observation in supervised learning issues contains variable of observed output and more than one observed variable of input. The definition of set of tests is a formation of data used to access the module efficiency using a metric of performance. It's difficulty that that none of the observations from the set of training compose into the set of tests. It will be difficult to tell if the algorithm has learned to generalize from the training set or has just memorised it if the test set contains examples from the training set. A software that generalises successfully will be able to fulfil tasks with fresh data effectively.

Over-fitting is the practise of memorising the workout set. A software that memorises its observations may not perform well because it may memorise noise or coincidental relations and structures. During the testing stage, the CNN style receives a picture sequence from the test and used for final trained network weights to predict the expression of each frame. We consider a person to be stressed if the anticipated images are stress associated facial expression.

5.3 Conclusion

A facial expression identification model is related to a convolution neural network, and a module for detecting negative emotional stress. The findings of this study showed that it was possible to detect acute cognitive stress in real time using CNN, and that it had practical implications. This is a difficult problem that has been tackled previously using a variety of methods. While feature detection has yielded positive results, this effort focused on feature learning, which is one of the DL techniques. We will look into applying the transfer depiction gained out of the face identification to face expression in future.

References

[1] Karpathy, A, Toderici, G., Shetty, S., Leung, T., Sukthankar, R. and Fei-Fei, L. (2014). Large-scale video classification with convolutional neural networks. *2014 IEEE Conference on* Computer Vision and Pattern Recognition (CVPR), 1725-1732. Doi: 10.1109/CVPR.2014.223.

[2] Pediaditis, M., Giannakakis, G., Chiarugi, F., Manousos, D., Pampouchidou, A. Christinaki, E., Iatraki, G., Kazantzaki, E., Simos, P. G., Marias, K., Tsiknakis, M. (2015). Extraction of facial features as indicators of stress and anxiety. 2015 37th Annual International Conference of the IEEE Engineering in Medicine and Biology Society, 3711-3714. Doi: 10.1109/EMBC.2015.7319199

[3] Andrychowicz, M., Denil, M., Colmenarejo, S. G., Hoffman, M. W., Pfau, D., Schaul, T., Shillingford, B., and Freitas, N. D. (2016). Learning to learn by gradient descent by gradient descent. 30th Conference on Neural Information Processing Systems (NIPS 2016), Barcelona, Spain, pp. 1-9.

[4] Mal, H. P., Swarnalatha, P. (2017). Facial expression detection using facial expression model. *2017 International Conference on Energy, Communication, Data Analytics and Soft Computing (ICECDS)*, 259-1262. Doi: 10.1109/ICECDS.2017.8389644.

[5] Li, H., Li, G. (2019). Research on Facial Expression Recognition Based on LBP and deep learning. *2019 International Conference on Robots & Intelligent System (ICRIS)*, 94-97. Doi: 10.1109/ICRIS.2019.00032.

[6] Zhang, J., Mei. X., Liu, H., Yuan, S., Qian, T, (2019). Detecting

Negative Emotional Stress Based on Facial Expression in Real Time. *2019 IEEE 4th InternationalConference on Signal and Image Processing (ICSIP)*, 430-434. Doi: 10.1109/SIPROCESS.2019.8868735.

6

Keeping Cyber-Physical Systems Secure Using Neural Network and Deep Learning

Manisha Verma
Shree Dhanvantary College of Engineering and Technology, Surat, Gujarat, India.

Abstract: Cyber-physical systems (CPS) are growing at good pace. Their use in different applications has been increasing. Their demand and popularity is growing. In such case, keeping them securing them is important. It needs to be made sure that they are secure. In this chapter, an introduction of CPS will be provided. Along with it, how CPS can be kept secure using neural network and deep learning will also be discussed.

6.1 Overview of Cyber-Physical Systems (CPS)

CPS is a technology which is presently trending across various sectors. CPS contains of physical components as well as computational components which are both working together for implementing a process in real time [13].

CPS are systems which combines capabilities of communication, computing and data storage, with the purpose of controlling or monitoring the entities which exist in the physical world [17].

CPS has the ability to integrate information and integrate operational technologies in terms of embedded, control and physical systems for forming new functionalities or improved functionalities [16].

Bringing an increase in levels of integration and automation across the design-operation time continuum, so called DevOps is included in common trends for CPS. This positioning of CPS's this positioning gives opportunities which are unprecedented for innovation across and within existing domains [16].

There is expectation from CPS that it will get revolution in virtual world and physical world interaction, similar to how interpersonal communication and interaction has been revolutionized by the internet [15].

CPS is known as realization technology. CPS creates new reality space with different processes and applications that are innovative and they dissolve the boundaries between real space and virtual space [15].

CPS has many different characteristics. Some of the characteristics of CPS are as follows:
- Robustness
- Autonomy
- Efficiency
- Communicative, etc [13].
- Robustness: CPS needs to be robust so that it can tolerate any system disturbances.
- Autonomy: CPS needs autonomy as a CPS should possess decision making capability.
- Efficiency: Efficiency is one of the most important and most needed characteristics of CPS.
- Communicative: CPS needs to be communicative so that data can be communicated and shared with the needed systems.

CPS can be used in different applications. Some of its applications are as follows:
- Smart buildings
- Smart factories
- Automated vehicles
- Automated personalized healthcare, etc. [13]

CPS provides a lot of advantages. Some advantages of CPS are as follows:
- Better System Performance
- Scalability, etc. [14].
- Better System Performance

With cyber infrastructure's and sensors close interaction, CPS can provide system performance which is better when it comes to feedback and redesigning automatically.
- Scalability

Different resources can be provided to the users as per their requirement by CPS as part of cloud computing [14].

6.2 Literature Survey

The authors in this paper have explained a particular way approach for detection of cyber attacks in CPS. Recurrent neural network is useful in the approach and specifically long short term memory recurrent neural network (LSTM-RNN) plays an important role. A learning approach which is unsupervised and is utilizing a RNN which is a time series predictor as their model is described and explained by the authors. A water treatment plant's similar version is considered and to find out anomalies in it, the authors then utilize the cumulative sum method [12].

The authors suggest using LSTM-RNN for data sequence prediction for detection of anomaly. The authors in the form of a predictor for modeling normal behaviour use LSTM-RNN. After that cumulative sum method is used for identifying behaviour which are abnormal [12].

This method which is proposed by the authors, along with detecting anomalies in CPS, also finds out the sensor that was attacked. With the help of Secure Water Treatment Testbed (SWaT), complicated dataset was gathered [12].

Different experiments were carried on this dataset which is complicated. With the help of the experiments, the authors show that most of the attacks which were designed by the authors research team can be detected with low false positive rates [12].

There has been growth in attack surface of industrial CPS's but resisting different cyber related threats for industrial CPS's is difficult. The authors in [11], have proposed DeepFed scheme which is a federated deep learning scheme. It helps for detection of industrial CPS's cyber threats [11].

Utilizing convolutional neural network (CNN) and a gated recurrent unit, the authors for industrial CPS design an intrusion detection which is deep learning based. Then authors in [11] develop a federated learning framework, enabling more than one industrial CPSs in a privacy-preserving manner to together make a intrusion detection model which is comprehensive [11].

In order to maintain privacy and security of model parameters via training process, a secure communication protocol is crafted which is based on Paillier cryptosystem. The proposed scheme, which is the DeepFed scheme, its high effectiveness for detecting different types of cyber threats to industrial CPSs was demonstrated by large experiments on a dataset of real industrial CPS [11].

The authors in [3], on the basis of machine learning (ML) models, presented a novel anomaly detection method. In this method, a genetic algorithm optimizes forecasting algorithm automatically. For predicting actuator readings and sensor readings, CNN is used in the framework proposed and this framework for measuring deviations from normal system behavior, uses errors it got from difference between the real truth of the ground and values that were predicted [3]. The authors in [3] for this task, compared three different methods performance. These methods were:

- K-means clustering
- An engineered threshold which are based on the errors of train set
- SVM [3].

The authors obtained an F1-score of 87.89%, with their proposed approach [3].

The authors inspect the attack detection and attack identification which is done on the basis of deep learning technology on wheel speed sensors of automotive CPS. A specific value is estimated in order done to substitute data which is false for quicker recovery of physical system along with cyber-attacks detection [9].

The authors, in this work [9], for improving safety of CPS including the ones which are even under the attacks, the authors, design a novel method for combining detection of attack. It even identifies and estimates vehicle speed of wheel speed sensors [9].
Based on the attack cases which can occur, the authors in [9] follow the below steps:
- First, the authors define states of the sensors.

- Then RNN is applied for detecting and also identifying wheel speed sensor attacks.
- After that, the authors employ Weighted Average (WA) is employed by the authors. It is employed to provide each sensor with different weight [9].

Experiments are conducted with real measurements which were obtained during driving on actual road. Classification accuracy is evaluated, in the case of detection and identification of fault [9].
For verifying speed estimation is done accurately, mean squared error (MSE) calculation is required to be carried out which is done by the authors in [9].

The authors demonstrate that integrity of the data is well maintained by their system. They also demonstrate that their system is relatively safe when compared with systems that apply other algorithms [9].

6.3 Discussion

In the literature survey, we discussed and saw different papers in which the authors propose and demonstrate various methods and schemes to keep CPS secure. Each of the methods discussed in the literature survey seem to be helpful in keeping CPS secure. Overall, it is seen that neural network, deep learning, machine learning actually helps in securing CPS.

6.4 Conclusion

In this chapter, an overview of CPS is provided. Also, how CPS can be kept secure is briefly discussed. In the recent years, applications of CPS and its popularity has been growing. However, there is a concern that how it can be kept secure. To take care of this concern, different methods are proposed and demonstrated by various authors as seen in literature survey in this chapter. These methods included the use of neural network and deep learning. It was seen that neural network and deep learning could very well help in keeping CPS secure.

References

1. Lee, E. A. (2010). Cps foundations. Proceedings of the 2010 Design Automation Conference. 737–742, doi: 10.1145/1837274.1837462.
2. Lee, E. A. (2008). Cyber Physical Systems: Design Challenges. *2008 11th IEEE International Symposium on Object and Component-Oriented Real-Time Distributed Computing (ISORC)*, 363–369, doi: 10.1109/ISORC.2008.25.
3. Jensen, J. C., Chang, D. H. and Lee, E. A. (2011). A Model-Based Design Methodology for Cyber-Physical Systems. *2011 7th International Wireless Communications and Mobile Computing Conference*, 1666-1671.
4. Kim, K-D., and Kumar, P. R. (2013). An Overview and Some Challenges in Cyber-Physical Systems. *Journal of the Indian Institute of Science*, 93(3), 341–352.
5. Perales Gómez, Á. L., Fernández Maimó, L., Huertas Celdrán, A., García Clemente, F. J. (2020) MADICS: A Methodology for Anomaly Detection in Industrial Control Systems. Symmetry *12*(10), 1583, doi: https://doi.org/10.3390/sym12101583
6. Rajkumar, R., Lee, I., Sha, L., Stankovic, J. (2010) Cyber-physical systems: The next computing revolution. *Design Automation Conference*, Anaheim, CA, USA, pp. 731-736, doi: 10.1145/1837274.1837461.
7. Lee, E. A. (2015). The past, present and future of cyber-physical systems: A focus on models. Sensors, 15(3), 4837–4869.
8. Kriaa, S., L. Pietre-Cambacedes, Bouissou, M., and Halgand, Y. (2015) A survey of approaches combining safety and security for industrial control systems. *Reliability engineering & system safety*, 139, 156–178.
9. Shin, J., Baek, Y., Lee, J. Lee, S. (2019) Cyber-Physical Attack Detection and Recovery Based
on RNN in Automotive Brake Systems. *Applied Sciences*, 9, 82, 1-21.
10. Oliveira, N., Sousa, N., Oliveira, J. Praça, I. (2021). Anomaly Detection in Cyber-Physical Systems: Reconstruction of a Prediction Error Feature Space. *2021 14th International Conference on Security of Information and Networks (SIN)*, Edinburgh, United Kingdom, 1-5, doi: 10.1109/SIN54109.2021.9699339.
11. Li, B., Wu, Y., Song, J., Lu, R., Li, T., Zhao, L. (2021) DeepFed: Federated Deep Learning for Intrusion Detection in Industrial Cyber–Physical Systems. *IEEE Transactions on Industrial Informatics*, 17(8), 5615-5624, doi: 10.1109/TII.2020.3023430.

12. Goh, J., Adepu, S., Tan, M., Lee, Z. S. (2017) Anomaly Detection in Cyber Physical Systems Using Recurrent Neural Networks. *2017 IEEE 18th International Symposium on High Assurance Systems Engineering (HASE)*, Singapore, 140-145, doi: 10.1109/HASE.2017.36.

13. Lozano, C. V., Vijayan, K. K. (2020) Literature review on Cyber Physical Systems Design. *Procedia Manufacturing*, 45, 295-300.

14. Bhrugubanda, B. (2015) A Review on Applications of Cyber Physical Systems. *International Journal of Innovative Science, Engineering & Technology*, 2 (6), 728-730.

15. Kim, SH. (2017) CPS (Cyber Physical System) based Manufacturing System Optimization. *Procedia Computer Science*, 122, 518-524.

16. Törngren, M., Sellgren, U. (2018) Complexity challenges in development of cyber-physical systems. In: Marten Lohstroh, Patricia Derler, Marjan Sirjani (ed.), *Principles of modeling:*
Essays dedicated to Edward A. Lee on the occasion of his 60th birthday, pp. 478-503. Switzerland: Springer

17. Rawung, R. H., Putrada, A. G. (2014) Cyber Physical System: Paper Survey. *2014 International Conference on ICT For Smart Society (ICISS)*, Bandung, Indonesia, pp. 273-278, doi: 10.1109/ICTSS.2014.7013187.

7

Intelligent Systems in Biomedical Domain: The Perils and Promise of Hope

Kalpana and Abhishek Maurya
Department of Biotechnology, Dr. Ambedkar Institute of Technology for Handicapped, Kanpur, UP, India

Abstract: Intelligent systems based on Artificial Intelligence (AI) and Machine Learning (ML) are poised to make disruptive and transformative advances in the biomedical domain. These systems assist in contexed-relevant data synthesis and automation in the biomedical industry. A formal definition of AI states 'the capability of a machine to imitate intelligent human behavior.' The AI-laden systems assist humans in modeling human reasoning to execute a problem or bypassing human logic and exclusively use a large volume of information to engender a solution. On the other hand, the AI system assimilates elements of human reasoning without accurate modeling of human processes. The framework of the AI system consists of two subdomains -methods and application. The Methods subdomains include evolutionary computing, expert systems, machine learning (ML), fuzzy systems, and probabilistic methods. The neural networks and support vector machines are two subdivisions of ML. The probabilistic methods are subdivided into the Bayesian networks and Hidden Markov models (HMM). The application domains include Natural Language Processing (NLP), predictive analytics, robotics, vision (Image recognition, machine vision), text-to-speech, and speech-to-text.

Symbolic AI is based on the high-level human-readable representation of problems and logic. ANNs are nowadays provide reliable results related to biomarker identification and classification of disease. With the help of artificial intelligence, it is possible to collect more information and samples on digital platforms, store data, and perform data analytics to pile up the affluence of data from biomedical databases such as genetic mapping on DNA sequences. The classification of genes and cancer cells, protein function, disease diagnosis, and disease treatment is more accurately assisted via AI implementation. Bio computation systems help improve various fields such as molecular medicine, drug development, and gene therapy. This chapter explores the potential of AI and ML in the biomedical domain. Intelligent systems can be made using AI and ML for application in biomedical domain. This chapter covers an overview of machine learning, artificial neural networks (ANNs) and AI. Also, the application of AI and ML in the biomedical domain is discussed in the chapter.

7.1 Introduction

Artificial intelligence (AI) is a comprehensive branch of computer science dedicated to creating intelligent machines that can accomplish activities that usually demand human intelligence. The four types of AI are the theory of mind, self-awareness, limited memory, and reactive devices. The reactive AI devices have no memory and are only used for a single task. The theory of mind is a psychological concept and, when implemented in machines, makes their social intelligence understand human emotions. The AI devices based on limited memory can draw inferences from prior experiences to help them make better decisions in the future. The devices become capable of predicting and inferring human behavior. The AI system having self-awareness is the future process for creating self-conscious machines. The typical examples of machines/devices supplemented with AI are Alexa, Siri, self-driving cars, etc. AI systems, in general, work by swallowing huge volumes of labeled training data, analyzed the data for patterns and correlations, and forecast solutions. Learning, reasoning, and self-correction are three cognitive skills for AI development. The learning element of AI programming involves collecting the data and constructing rules for turning it into a useful information set. Algorithms are rules that give machines step-by-step instructions for completing a particular task. The reasoning process of AI is concerned with picking the best algorithm to achieve a conclusive output. Self-correction strategies are concerned with fine-tuning algorithms and verifying that they deliver the most precise solutions possible.

From the inception of AI in the early 1950s to the 1990s, the general approach of AI was Symbolic AI. The foundation of symbolic AI was laid on programmers' complex coding protocols for executing complex tasks through machines. Initially, this approach worked well but became unsuccessful in the execution of tasks that required human intelligence. Here a subset of AI which is Machine Learning (ML) can be helpful. The ML approach flattered the flame of AI machines. IBM developed an AI-supported Deep Blue machine. A chess competition was held in 1994 between Deep Blue and Garry Kasparov, a reigning chess champion. The competition was won by Deep Blue and flagged the initial victory of AI.

Deep Blue executed the decision-making task during the game, but human assistance was obligatory for the physical movement of the chess pieces. It was a critical example of exploring the interconnection between human intelligence and AI by examining the dissimilitude between physically moving the pieces and decisions about chess moves. Although chess is considered an obstinate game but its rules are very elementary. The Deep blue was programmed to compute a set of ahead likely actions rendering on regulations, and the machine just opted for the best action during the game. This was the remarkable success story of reactive machines at that time. It was also a typical example of a Symbolic AI approach application. Now let us explore the other facet of the model as mentioned above. The coding program executes the moving of pieces based on rules, but what laws would have to do with selecting a chess piece? Human perception begins with recognizing the pieces' current location or, in machine terminology, image recognition. For this, any rules or program will allocate the machine for image recognition. The complexity of achieving these tasks is the cause for the failure of symbolic AI to execute easy tasks such as image recognition over enormous achievements in complex tasks such as chess moves execution. Deep learning systems have been applied widely and set new benchmarks in the economy where significant advanced data are abundant. There is a solid financial motivating force to robotize expectation tasks.

Similarly, the explicit-coded rules of symbolic AI do not reinforce any similarity to the fuzzy nature of human intelligence. At this point, Machine Learning (ML), a subset of AI, enters the machine world. The Machines laden with ML learn to solve tasks themselves instead of being given a set of explicit rules on how to resolve the given problem. It is usually attained by assimilating new practices and positive feedback gained through repeated tasks. It is the foundational similarity between human intelligence and AI. Machine learning applications are predominantly classified into two categories based on whether a 'feedback (unsupervised learning)' or 'label (supervised learning)' learning approach is available to machines. Both supervised and unsupervised learning approaches have tremendous applications as well as future potential in the biomedical domain. The recognition of patterns, text interpretation, and object detection are some common application areas of ML. The ML also facilitates machines for learning and improvement automatically from experience without being programmed explicitly. The foremost objective of machine learning is to finding out the patterns and making assumptions that depend upon the complex design of data. Supervised learning, semi-supervised learning, unsupervised learning, and reinforcement learning are four basic approaches to ML. The new approach in AI is due to one group of procedures, specifically, deep learning. The deep learning approach customized machines to learn affiliations dependent on enormous amounts of crude information like the pixels of digital images. The industry is putting resources into the innovation alongside high execution processing abilities, improved capacity, and equal figuring infrastructure, its applications in our day-to-day routines have made it progressively significant for us. The most straightforward application of AI techniques in the biomedical field includes assessing and analyzing patient records, and clinical informatics. Another down-to-earth application is noticed in regions like liver pathology, thyroid sickness determination, rheumatology, cardiology, neuropsychology, etc. [4]. Clinical informatics is currently confronting genuine mishaps as we have not had the option to prepare measurable strategies skilled in managing loud and missing information. Because of this reason, the outcomes coaxed out of an AI probe clinical information face vulnerability and blunders [18]. The developing pattern in web world has concocted a moving new framework which is known as the Internet of things (IoT), wherein a few gadgets are interconnected and continue to share valuable tangible information, helping gadgets comprehend and react to the different outside aspects. The strategy is presently making new openings in a broad scope of areas like medical care, retail, banking, producing, and customized client applications [34].

7.2 Machine Learning

The ML is becoming popular as its algorithm help recognize pattern, text interpretation, and object detection. It mainly focuses on the data uses and designing algorithms that mimic what a normal human being can learn. It gently enhances its perfection. It provides the capability to the system that it can understand and improve automatically from experience without being programmed explicitly. As the field of data science is developing, machine learning has become the critical component.

Machine learning employs an extensive pair of statistical techniques compared to other approaches used in medicines. The objective behind ML is to find out the data patterns and make assumptions that depend upon the complex design to answer business questions.

Supervised learning, reinforcement learning, unsupervised learning, are different components of ML.

7.2.1 Decision Tree

The decision tree is a machine learning technique used to categorize common traits. A decision tree constitutes a rule which toils for classifying data according to these traits. A decision tree comprises nodes, leaves, and edges. The node assists in specifying the features for data separation. Each node consists of several edges. An edge connects two nodes or a node and a leaf depending on the tree's function. The leaves are labelled with decision values for data classification.

A decision tree is used to detect intrusive behaviour, which means the output and corresponding output are as per the training data. Here, the data is splitting continuously with parameter changes. A Decision tree copies the thinking ability of a human being while making a decision and making it easy to understand for others.

7.2.2 Genetic Algorithms

The Genetic Algorithm (GA) is a subset of evolutionary algorithms. The GA deploys the concept of natural selection and genetics for solving the problems related to optimization in lesser time. It is helpful in various real-time applications like designing electronic circuits, code-breaking, and data centres. The notion of GA mimics the process of natural selection and survival of the fittest. The operational procedure of GA is based on five phases-
   i.   Initial population

ii. Fitness function-- New chromosomes are evaluated for their fitness.
iii. Selection- The best pair of the gene is selected upon the fitness function is chosen.
iv. Crossover - New offspring are generated by the process of recombination and mutation.
v. Mutation- this is the last step; in this step, the old population is replaced with the newly generated population.

The initial population represents a set of individuals or a group of solutions for the given problem. Each individual possesses a set of variable parameters or genes. The Genes are arranged on a string to form a solution or chromosomes. The coding unit is termed a gene, and the sequence which is going to be encoded is termed as chromosomes. The fitness function evaluates the fittest score of an individual in a population. The fitness score determines the probability of selection of an individual for reproduction. The GA starts with the randomly created population. In the next step, we calculate the fitness function of each chromosome then repeat the procedure until the "n" offspring are made in the selection stage. The GA also predicts and supervises the activity of chromosomes and changes in genetic structure.

In GA, it starts with the chromosome's population and evaluation which measures the ability of the chromosomes. For making a new solution, it employs reproduction and mutations. This process of recombination and evaluation was repeatedly performed several times. If the problem is constructed reasonably, the solution strongly emerges gradually.

7.2.3 Artificial Neural Network (ANN)

The ANN or neural network, or simulated neural networks (SNNs) are a subset of ML. The ANNs acts like neurons for signal transduction. The neurons are synchronized in node layers, and output signals are passed from one neuron to others present in the next layer. The neuron layers consist of the input layer, an output layer, and one or more hidden layers. The hidden layers increase the computation power of the neural network. The individual neurons in the network are assigned with a threshold and weightage value. The node becomes activated and transfer data to the next layer of the network when the output value is above the threshold. The flow of information is always unidirectional and in a layer-by-layer sequence. A training set of data is required for neural network training and to improve network accuracy over time. Though these ANN algorithms are trained for accuracy, the trained ANN algorithms are indispensable tools of AI for data clustering and classification at high speed. The google search algorithm is one of the most famous examples of ANNs.

The most straightforward neural network system is called perceptron [5]. This neural network is a unit that receives information from various branches. It is a single neuron classifier; with the help of the threshold activation function, the linear discrimination function separates into two classes.

7.2.4 Support Vector Machine

Super vector machine is a specified machine learning algorithm that is primarily employed for both the classification and regression challenges. In this algorithm, a plot is there for each data concerning 'n' dimensional space. Here 'n' represents the number of features, with each feature value being the value of a specific coordinate.

The training pattern of the support vector is identically close to the hyperplane. The supporting vectors are training samples that specify hyperplane separating optimally, which was not easy to classify. The separation can reformulate the minimizing problem of the magnitude of the mannered weight vector.

Support vector is characterized as two types:
- Linear Support Vector
- Nonlinear Support Vector

7.3 Probabilistic Method of Machine Learning

Probabilistic methods of ML give a set of tools for representing uncertainty, performing inference, and making decisions. The two main methods of probabilistic machine learning are the Bayesian network and Hidden Markow Model (HMM).

7.3.1 Bayesian Network

The Bayesian method of machine learning authorizes estimating the uncertainty in prediction, which is an advisable feature in this field. The goal of Bayesian machine learning is to evaluate the posterior distribution ($p(\theta/x)$). The construction of a statistical model based on Bayes's theorem shows that Bayesian machine learning is a paradigm.

Bayesian machine learning aims to estimate the posterior distribution that is ($p(\theta/x)$). In recent years Bayesian networks have been growing in many books and theoretical and practical publications. In the highly score + search approaches, the search is directed in the acyclic graph that represents a feasible Bayesian structure [29].

7.3.2 Hidden Markow Model (HMM)

It is a probabilistic method of machine learning. Primarily HMM is used in speech recognition and sometimes for the classification task. The HMM is regarded as the most critical machine learning model, especially in speech recognition and processing language, recognizing sign language, and handwriting.

Markov chain and Hidden Markov Model both are the finite automata extension depends on the observation provided by input. The assumption is taken by Markov after specifying the Markov chain by using certain components like:

$Q = q_1 q_2 .... q_n$ termed as a set of states
$A = [a_{ij}]_{NXN}$ A is the transition probability matrix, a_{ij} states the moving state probability of i to j.
q_0, q_{end} represent the start and end state.

Overall Markov Assumption: $P(q_i/q_1....q_{i-1}) = P(q_i/q_{i-1})$

$P(q_j/q_i)$ states the law of probability and shows the value of probability must be 1.

The Markov chain is helpful to compute the sequence probability of an event that the world can observe and those events in which the only one is interested but may not be directly observable in the world.

7.3.3 Fuzzy systems

Fuzzy logic is a technique that is utilized for data handling purposes. It permits ambiguity and is predominantly used in the medical field. It finds out about fuzziness in a computationally proficient manner. This strategy is utilized in numerous medical areas like multiple logistic regression analysis and used for the diagnosis of multiple diseases like lungs cancer, acute leukaemia, etc.

. Intelligent systems can be made using AI and ML for application in biomedical domain as mentioned earlier. First we will discuss applications of AI in biomedical domain.

7.4 Applications of AI in Biomedical Domain

7.4.1 Genomics and AI

AI has made tremendous progress in computer science throughout the most recent ten years and is still among the quickest developing regions. By 2014, mainstream researchers had published various research works where AI is applied to explain genome information retrieval and management. In any case, the wide-scale down-to-earth application is something yet to occur. Genes understanding would assist with changing medication across the globe [25]. A lot of training can be given to computers and this its advantage.

Examining biological sequence data such as viral genomic and proteomic sequences requires either conventional AI or progressed deep learning, or a combination of both depending upon issues being tended to and data pipelines utilized.

AI-driven algorithms assist in recognizing the locations of starting sites of transcription, splice site identification, pinpointing promoters, etc. in a genome sequence.

7.4.2 Proteomics and AI

Proteins came into the picture when acquiring them in purified structure utilizing Mass Spectroscopy and Blotting approaches. Since then, the improvement of high-throughput strategies in protein-based investigations, likewise called 'proteomics,' has been growing. With more data accessible, AI has discovered expanded applications in the forecast, including determination, pattern recognition, and various robotization works. The significant application is as semi-supervised learning methods where the algorithms learn from enormous datasets out of which just a few are labeled.

7.4.3 AI and Proteome Informatics

Protein structure determination and dynamic examination of the predicted structures are the absolute first regions to apply AI as decoding of protein structures is fundamental for the comprehension of biological cycles and understanding cell working [10, 37].

Machine learning-based AI is effectively being applied in protein fold expectation and design forecast applications.

7.4.4 AI in Biomedical Research

Recently, biomedical research has turned into a data-driven activity empowered by book material, and test rehearses connected to data collection, distribution, and use.

AI is also set to produce visionary models that could help doctors/specialists in prognostic appraisal and personalizing treatment and rehabilitation for individual patients, for example, in the aftermath of a stroke. Electronic health records (EHR), for instance, offer the chance to utilize genuine world data to generate information about the results of a given medical procedure (be it analysis, a prognosis, a therapy, or a rehabilitation plan [41]. Artificial intelligence can be utilized to mine EHR to find disease commonality or individuals in danger for a given chronic disease and work on the association of health frameworks by offering help in emergency and patient administration. In a new report, deep learning was utilized to make prescient demonstrating with EHR to precisely measure in-clinic mortality, readmission odds, length of stay, and final discharge diagnoses [36]. In another examination, an AI algorithm distinguished cancer patients at danger of 30-day mortality before they start chemotherapy (both palliative and curative) [18]. Such an algorithm can help choices about chemotherapy commencement, allowing more judiciously allotment of assets [28].

Another worry identifies the adequacy of informed assent as a moral defence in research, including algorithmic processing. The customary idea of informed assent is already tested in instances of data gathered in more traditional exploration settings, as it is increasingly difficult to foresee who will get to the data later on, for which purposes, and under which conditions. The endless uses of data and the linkage of different data collections makes even the thought of wide assent difficult to expect. On account of AI, it is as yet not as satisfactory due to certain issues like data privacy and security.

Artificial intelligence adds a layer of moral intricacy in that it utilizes data to separate fine-grained information about people. Researchers have a moral obligation to safely shield this information from unapproved access to avoid security-related damages to data subjects throughout exploration projects. The undesirable break of heath applicable information can prompt discriminative employments of such information in spaces like employment, education, and insurance. This issue applies to data produced and put away by researchers and reports that researchers input to investigate members as primary, secondary, or incidental discoveries [9].

One more issue of moral pertinence regarding health research has risen out of collaborations between companies with advanced abilities in AI and medical care organizations in control of health data indexes. While such coordinated efforts can be helpful together, there can be concerns with respect to data security and as well as in other areas. All the concerns need to be evaluated and taken care of.

7.4.5 AI and Medicine

In the advanced world, medicine assumes a significant part in the existence of humanity. Because of advances in medicine, you can accomplish a phenomenal future and work on its quality. Nonetheless, a sharp expansion in the measure of clinical information has prompted the requirement for superior grades and quick preparation. Artificial intelligence and machine learning frameworks should help in reducing clinical error issue. Another factor that makes AI frameworks amazingly encouraging is the relative expense viability and advantages of utilizing these frameworks.

Presently, AI is characterized as different programming frameworks and the method and algorithms utilized in them, the fundamental component of which is the arrangement of issues, similar to an individual.

Artificial intelligence (AI) permits systems to gain from their insight, adjust to set boundaries, and perform tasks that were already just humanly conceivable. Because of these advancements, systems can be taught to play out specific undertakings by handling a lot of information and distinguishing pattern in them. In most AI executions, from PC chess players to driverless vehicles, the capacity to learn deep and process natural language is critical.

Expert systems are applied AI frameworks in which the information base is a formalized practical information on exceptionally qualified trained professionals (specialists) in a limited branch of knowledge. Expert systems (ES) are intended to supplant specialists when tackling issues because of their inadequate number, lack of productivity in taking care of the issue, or dangerous (harmful) conditions. Typically, expert systems are considered according to the perspective of their application in two angles: for what undertakings they can be utilized and in what field of activity. These two perspectives transform the design of the expert system being created. The accompanying primary classes of assignments that expert systems can settle can be recognized: diagnostics; forecast, identification; management; design; observing.

The development of an ES is conceivable just in case there are specialists in the field, and the specialists should concur in their evaluation of the proposed arrangement; the issue should have a place with a sufficiently structured area; the arrangement must not utilize a lot of sound judgment (i.e., a wide scope of general information about the world and how it functions), however, should be founded on some information to infer target information.

The exchange of information starting with one expert person then onto the next is troublesome, not normal for the exchange of data between ES. This is a basic course of duplicating information starting with one framework then onto the next, without the need to re-lay the data and long-term preparing.

This segment of AI theory spotlights discovering techniques for taking care of issues by figuring out how to tackle the comparative problems. One of the principal objective of AI in biomedicine is to help the clinical personnel and the patients in best way possible.

7.4.6 Bio computation and Cancer treatment

When seen as data advancements, it is prominent biological frameworks have a gigantic capacity as control systems for deftness or guideline, for pattern recognition, adaptability, data storage, sensor fusion, and other data taking care of assignments of extraordinary interest to computers researchers, computer specialists and any other individual keen on IT-related research. In these and different spaces, science performs at levels many significant degrees better than silicon-based frameworks [24]. We accept that the biology and information technologies interface exploration might prompt significant new information frameworks (algorithms and software) and computer advancements (hardware) [26]. The inquiry is the thing that and how might we take in and comprehend from the biological frameworks, and how might we take on them and adapt them to develop these new computer innovations. This short composition aims to characterize what we mean by Biological Computation, taking into account other work around here and afterward develop the plan to fill in as a reason for future conversation [11].

Bio computation fills in as a numerical device to analyses the fundamental physical principles that influence the conveyance and degradation of nanoparticles in cancer treatment. It gives a superior comprehension of the connection between nanotechnologies and living tissue subsequently saves time and assets by providing a reasonable structure to anticipate trial results. The significant bio computation test is to fit these basic standards into a biologically relevant model for the specific bio-numerical product. It is hard to set up a model from nano molecule to growth scale, not just for the explanation that matter acts contrastingly in each, somewhat because reproduction might require reconciliation of numerous chains of command of models, each varying in a few sets of extent as far as scale and subjective properties.

In cancer therapy, it is trusted that bio computation will ease the plan of ideal treatment models that would permit organization procedures for chemotherapy to take advantage of advantage while downestimating the side effects [17, 27].

Bio computation-based theoretical outcomes might actually be approved by the relationship of numerical prediction with in vitro and in vivo data of a specific patient's cancer response to chemotherapy. Thus, these tentatively and clinically approved bio computation results might be utilized to configuration customized treatment in silico utilizing computer simulations [39]. Since there are no incorporating numerical models that can apply to all possible physical and chemical measures being used for improvement of a particular medical product, it is important to develop a satisfactory theory considering the required medical treatment.

7.5 ML in Biomedical application

Neural networks depend on displaying measures that happen in the human cerebrum. Artificial neurons are joined in networks, associating the yields of specific neurons with the contributions of others. In working on terms, a neural network is a program that gets data at the info and offers responses at the yield.

There are likewise more complex models in which the yield of one network is coordinated to the contribution of another. These models make falls of neural networks, alleged multilayer neural networks.
Deep (machine) learning is a bundle of algorithms dependent on neural networks that endeavour to demonstrate significant level deliberations in information utilizing models comprising of numerous nonlinear transformations.

7.5.1 ANN application as a dynamic model, classifier and diagnosis tool for biomedical applications

In this paper, the authors [2] have used ANN in different ways, namely as a classifier, diagnosis tool and dynamic model. There are examples presented in the paper which include blood flow emboli classification. This classification is on the basis of transcranial ultrasound signals. Then there is tissue temperature modeling also presented in the paper which is based on imaging transducer's raw data and also mentioned is identification of ischemic cerebral vascular accident areas based on computer tomography images. In each of these mentioned cases, ANN performance is very good. It gives results which are accurate. The results encourage repeated use of these computational intelligent techniques on medical applications [2].

7.5.2 ANN for finding biomarkers

The accuracy of single biomarkers is not up to the mark. It is seen that rarely they have good accuracy. Suites of biomarkers can also have results which are conflicting. To identify specific analytes and their level of toxicity, potent combinations of variables which are ideal are isolated. Reduction of the thousands of candidate variables to something which is a small number that's required for treatment classification enables search for such combinations. When the key variables are recognized by machine learning (ML) Results which are received on recognition of important variables by ML are very surprising, given the other searching methods apparent failure to produce good diagnostics. For portable field tests of a variety of adverse conditions, proteins seem especially useful. This review by authors shows how ML, in particular artificial neural networks, can find potent biomarkers embedded in any type of expression data, mainly proteins in this article. A computer does multiple iterations to produce sets of proteins which systematically identify the treatment classes of interest. Whether these proteins are useful in actual diagnoses is tested by presenting the computer model with unknown classes [43].

7.6 Conclusion

AI and ML is being applied in biomedical and health care areas. Medicine has formed as an upscale testbed for AI and ML experiments discoveries inside the earlier decade, allowing researchers and developers to develop progressed and convoluted frameworks with the superpower of learning capacity. Presently, AI and ML-based frameworks are being worked upon in many applications in medicine area. This chapter has discussed and presented some applications of AI and ML in the biomedical and health care areas, along with an overview of AI and ML. The literature is introduced concisely and presents the comprehensive application of AI and ML in the biomedical domain health care space.

Acknowledgement

The authors are very thankful to Prof. Rachana Asthana, Director, Dr. Ambedkar Institute of Technology for handicapped for supporting us.

Conflict of Interest

The authors declare no conflict of interest.

References

1. Araghi, S., & Nguyen, T. (2021). A Hybrid Supervised Approach to Human Population Identification Using Genomics Data. *IEEE/ACM Trans. Comput. Biol. Bioinformatics*, *18*(2), 443–454. https://doi.org/10.1109/TCBB.2019.2919501
2. Ruano, M. G., Ruano, A. E. On the Use of Artificial Neural Networks for Biomedical Applications. In: Balas, V., Fodor, J., Várkonyi-Kóczy, A., Dombi, J., Jain, L. (eds) Soft Computing Applications. Advances in Intelligent Systems and Computing, vol 195. Springer, Berlin, Heidelberg. https://doi.org/10.1007/978-3-642-33941-7_40
3. Banerjee, A. K., Arora, N., & Murty, U. S. N. (2012). Aspartate carbamoyltransferase of Plasmodium falciparum as a potential drug target for designing anti-malarial chemotherapeutic agents. *Medicinal Chemistry Research*, *21*(9), 2480–2493. https://doi.org/10.1007/s00044-011-9757-3
4. Banerjee, A. K., Harikrishna, N., Kumar, J. V., & Murty, U. S. (2011). Towards Classifying Organisms based on their Protein Physicochemical Properties using Comparative Intelligent Techniques. *Applied Artificial Intelligence*, *25*(5), 426–439. https://doi.org/10.1080/08839514.2011.570158
5. Banerjee, A. K., Kiran, K., Murty, U. S. N., & Venkateswarlu, C. (2008). Classification and identification of mosquito species using artificial neural networks. *Computational Biology and*

Chemistry, 32(6), 442–447. https://doi.org/https://doi.org/10.1016/j.compbiolchem.2008.07.020
6. Banerjee, A. K., Ravi, V., Murty, U. S. N., Sengupta, N., & Karuna, B. (2013). Application of Intelligent Techniques for Classification of Bacteria Using Protein Sequence-Derived Features. *Applied Biochemistry and Biotechnology*, *170*(6), 1263–1281. https://doi.org/10.1007/s12010-013-0268-1
7. Banerjee, A. K., Ravi, V., Murty, U. S. N., Shanbhag, A. P., & Prasanna, V. L. (2013). Keratin protein property based classification of mammals and non-mammals using machine learning techniques. *Computers in Biology and Medicine*, *43*(7), 889–899. https://doi.org/https://doi.org/10.1016/j.compbiomed.2013.04.007
8. Chakraborty, I., & Choudhury, A. (2017). Artificial Intelligence in Biological Data. *Journal of Information Technology & Software Engineering*, *07*(04). https://doi.org/10.4172/2165-7866.1000207
9. Chakraborty, S., & Biswas, M. C. (2020). 3D printing technology of polymer-fiber composites in textile and fashion industry: A potential roadmap of concept to consumer. *Composite Structures*, *248*, 112562.
10. Cox, D. R. (2015). Big data and precision. *Biometrika*, *102*(3), 712–716. https://doi.org/10.1093/biomet/asv033
11. Cull, P. (2013). Biocomputation: Some history and prospects. *Bio Systems*, *112*(3), 196–203. https://doi.org/10.1016/j.biosystems.2012.12.005
12. Daszykowski, M., & Walczak, B. (2009). Density-Based Clustering Methods. *Comprehensive Chemometrics*, *2*, 635–654. https://doi.org/10.1016/B978-044452701-1.00067-3
13. Domshlak, C., Hüllermeier, E., Kaci, S., & Prade, H. (2011). Preferences in AI: An overview. *Artificial Intelligence*, *175*(7–8), 1037–1052. https://doi.org/10.1016/j.artint.2011.03.004
14. Duddela, S., Nataraj Sekhar, P., Padmavati, G. V, Banerjee, A. K., & Murty, U. S. N. (2010). Probing the structure of human glucose transporter 2 and analysis of protein ligand interactions. *Medicinal Chemistry Research*, *19*(8), 836–853. https://doi.org/10.1007/s00044-009-9234-4

15. Dunn, S. D., Wahl, L. M., & Gloor, G. B. (2008). Mutual information without the influence of phylogeny or entropy dramatically improves residue contact prediction. *Bioinformatics*, *24*(3), 333–340. https://doi.org/10.1093/bioinformatics/btm604
16. Elfiky, A. A., Pany, M. J., Parikh, R. B., & Obermeyer, Z. (2018). Development and Application of a Machine Learning Approach to Assess Short-term Mortality Risk Among Patients With Cancer Starting Chemotherapy. *JAMA Network Open*, *1*(3), e180926. https://doi.org/10.1001/jamanetworkopen.2018.0926
17. Feng, S.-S. (2011). Chemotherapeutic Engineering: Concept, Feasibility, Safety and Prospect—A Tribute to Shu Chien's 80th Birthday. *Cellular and Molecular Bioengineering*, *4*(4), 708–716. https://doi.org/10.1007/s12195-011-0198-3
18. Frankel, F., & Reid, R. (2008). Books & ArtSs Distilling meaning from data. *Nature*, *455*(September), 30.
19. Gandomi, A., & Haider, M. (2015). Beyond the Hype. *Int. J. Inf. Manag.*, *35*(2), 137–144. https://doi.org/10.1016/j.ijinfomgt.2014.10.007
20. Ghahramani, Z. (2015). Probabilistic machine learning and artificial intelligence. *Nature*, *521*(7553), 452–459. https://doi.org/10.1038/nature14541
21. Higuchi, N. (2013). Three challenges in advanced medicine. *Japan Medical Association Journal*, *56*(6), 437–447.
22. Ishida, T., & Kinoshita, K. (2008). Prediction of disordered regions in proteins based on the meta approach. *Bioinformatics*, *24*(11), 1344–1348. https://doi.org/10.1093/bioinformatics/btn195
23. Jacobs, A. (2009). The Pathologies of Big Data: Scale up Your Datasets Enough and All Your Apps Will Come Undone. What Are the Typical Problems and Where Do the Bottlenecks Generally Surface? *Queue*, *7*(6), 10–19. https://doi.org/10.1145/1563821.1563874
24. Kalpna, Srivastava, R. K., & Nath, R. (2018). Structure based drug designing against Inosine Monophosphate Dehydrogenase Receptor of Cryptosporidium parvum. In *International Conference on Bioinformatics and Systems Biology, BSB 2018*. https://doi.org/10.1109/BSB.2018.8770643
25. Karp, J. M., & Langer, R. (2007). Development and therapeutic

applications of advanced biomaterials. *Current Opinion in Biotechnology*, *18*(5), 454–459. https://doi.org/10.1016/j.copbio.2007.09.008
26. Katiyar, K., Kumari, P., & Srivastava, A. (2022). Interpretation of Biosignals and Application in Healthcare. In *Information and Communication Technology (ICT) Frameworks in Telehealth* (pp. 209–229). Springer.
27. Katiyar, K., Srivastava, R. K., & Nath, R. (2021). Identification of novel anti-cryptosporidial inhibitors through a combined approach of pharmacophore modeling, virtual screening, and molecular docking. *Informatics in Medicine Unlocked*, *24*, 100583. https://doi.org/10.1016/j.imu.2021.100583
28. Katiyar, S., & Katiyar, K. (2021). Recent trends towards cognitive science: from robots to humanoids. In *Cognitive Computing for Human-Robot Interaction* (pp. 19–49). Elsevier.
29. Larranaga, P., Kuijpers, C. M. H., Murga, R. H., & Yurramendi, Y. (1996). Learning Bayesian network structures by searching for the best ordering with genetic algorithms. *IEEE Transactions on Systems, Man, and Cybernetics - Part A: Systems and Humans*, *26*(4), 487–493. https://doi.org/10.1109/3468.508827
30. Leff, D., & Yang, G.-Z. (2015). Big Data for Precision Medicine. *Engineering*, *1*, 277. https://doi.org/10.15302/J-ENG-2015075
31. Lewis, N. E., Nagarajan, H., & Palsson, B. O. (2012). Constraining the metabolic genotype–phenotype relationship using a phylogeny of in silico methods. *Nature Reviews Microbiology*, *10*(4), 291–305. https://doi.org/10.1038/nrmicro2737
32. Libbrecht, M. W., & Noble, W. S. (2015). Machine learning applications in genetics and genomics. *Nature Reviews. Genetics*, *16*(6), 321–332. https://doi.org/10.1038/nrg3920
33. Malone, J. H., & Oliver, B. (2011). Microarrays, deep sequencing and the true measure of the transcriptome. *BMC Biology*, *9*. https://doi.org/10.1186/1741-7007-9-34
34. Perera, C., Ranjan, R., Wang, L., Khan, S. U., & Zomaya, A. Y. (2015). Big Data Privacy in the Internet of Things Era. *IT Professional*, *17*(3), 32–39. https://doi.org/10.1109/MITP.2015.34
35. Podell, S., & Gaasterland, T. (2007). DarkHorse: a method for genome-wide prediction of horizontal gene transfer. *Genome Biology*, *8*(2), R16. https://doi.org/10.1186/gb-2007-8-2-

r16.
36. Rajkomar, A., Oren, E., Chen, K., Dai, A. M., Hajaj, N., Hardt, M., ... Dean, J. (2018). Scalable and accurate deep learning with electronic health records. *NPJ Digital Medicine*, *1*, 18. https://doi.org/10.1038/s41746-018-0029-1
37. Schmidhuber, J. (2015). Deep Learning in neural networks: An overview. *Neural Networks*, *61*, 85–117. https://doi.org/10.1016/j.neunet.2014.09.003
38. Schmitt, R., Dietrich, F., & Dröder, K. (2016). Big Data Methods for Precision Assembly. *Procedia CIRP*, *44*, 91–96. https://doi.org/10.1016/j.procir.2016.02.141
39. Srivastava, A., Seth, A., & Katiyar, K. (2021). Microrobots and Nanorobots in the Refinement of Modern Healthcare Practices. In *Robotic Technologies in Biomedical and Healthcare Engineering* (pp. 13–37). CRC Press.
40. Suthaharan, S. (2014). Big Data Classification: Problems and Challenges in Network Intrusion Prediction with Machine Learning. *SIGMETRICS Perform. Eval. Rev.*, *41*(4), 70–73. https://doi.org/10.1145/2627534.2627557
41. Yager, J., Greden, J., Abrams, M., & Riba, M. (2004). The Institute of Medicine's report on Research Training in Psychiatry Residency: Strategies for Reform - Background, results, and follow up. *Academic Psychiatry*, *28*(4), 267–274. https://doi.org/10.1176/appi.ap.28.4.267
42. Yannakakis, G. N. (2012). Game AI Revisited. In *Proceedings of the 9th Conference on Computing Frontiers* (pp. 285–292). New York, NY, USA: Association for Computing Machinery. https://doi.org/10.1145/2212908.2212954
43. Bradley, B. P. (2012). Finding biomarkers is getting easier. *Ecotoxicology (London, England)*, *21*(3), 631–636.